JACKSONVILLE ON *Wheels*

JACKSONVILLE ON *Wheels*

A CAR CULTURE RETROSPECTIVE

DOROTHY K. FLETCHER

THE History PRESS

Published by The History Press
Charleston, SC
www.historypress.com

First published 2018

Manufactured in the United States

ISBN 9781625859433

Library of Congress Control Number: 2018948028

CONTENTS

PREFACE

I recall my so called misspent youth
It seems more worthwhile every single day
Cruisin' Van Nuys and acting so uncouth
All the joys of runnin' away
There was no speed limit on the Nevada state line
The air was red wine on those top down nights
Just you and me my old roller-skate
And the common sense to know our rights
Sweet old racin' car of mine
Roarin' down that broken line
I never been so much alive
Too fast for comfort
Too low to fly
Too young to die....
—Jimmy Webb, "Too Young to Die"

Many people who know me might be amazed to find that I have written a book about cars. I wasn't at all into "wheels" when I was growing up. I didn't even learn to drive a car until I was twenty-one. My insurance adjustor father had scared every car-loving molecule right out of me with his tales of car accidents he had to inspect, so I was content, for all of my younger days, to be a car passenger.

But now that I am much older and I think about my "misspent youth," I can remember some pretty spectacular "top-down nights" as a passenger flying down Mandarin's oak-lined roads or down A1A near any of North Florida's beautiful beaches. In living, it has become clear to me that cars are indeed an important part of life—my life. From the white 1970 VW Beetle filled with balloons my husband and I drove away after our wedding to the 2017 silver Passat that takes me all over Jacksonville, I have come full circle.

Cars have not only taken me from point A to point B. They have also provided me with countless fond memories and something even more intangible: freedom. They represent the freedom to travel long distances or just down to the corner store. Whether in sad times or joyous ones, whether for work or play, my automobile is almost a part of the family and is certainly a faithful friend.

I learned that lesson when a Subaru I once owned was stolen while I was at work back in the '90s. I was devastated at its loss, and I was certain I would never feel the same about the car, especially when it was found abandoned and gutted by the side of the road quite a distance away. Oddly enough, I was elated when they drove my shiny, repaired car out of the body shop. It

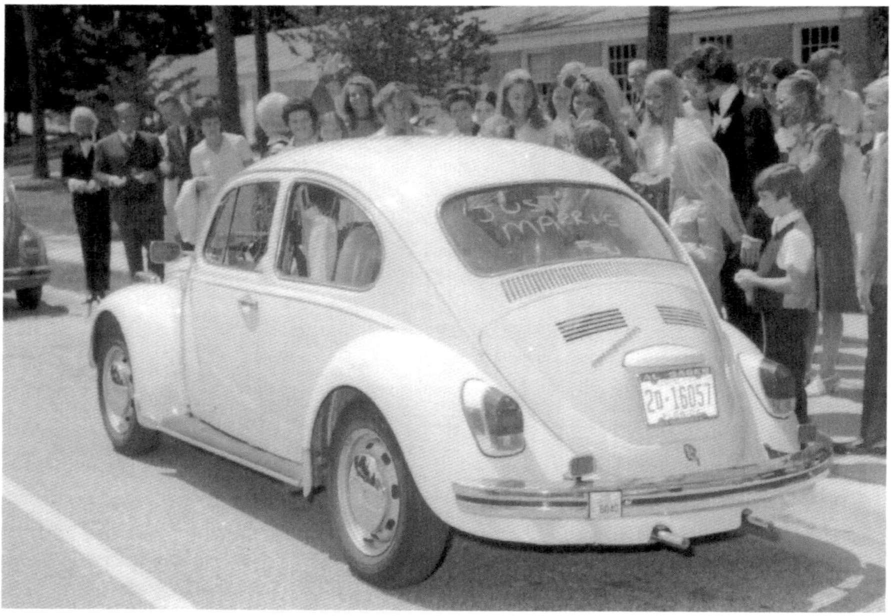

Wedding VW Beetle, 1971. *Author's collection.*

was as if a long-lost friend had moved back to town. Who knew that one could care so much for an inanimate object?

Of course, my husband Hardy's love of cars has helped in my gradual acceptance of car awesomeness. It started when he brought home a 1960 MGA to "fix up." We had been married for about a year, and I slowly warmed to the idea of this cute little car (which my father said would go right under a truck). Regardless of my father's misgivings, our MGA car was fun to drive and to ride in, and it lives fondly in my memory.

Hopefully my enthusiasm for all things automotive in this book will be as infectious to readers as well. Perhaps I can bring back some special memories of innocent, and not-so-innocent, times when we put the tops down and "roared down that broken line," back when there were so many roads left to travel.

IN DOING MY RESEARCH, I was surprised to learn that the *Florida Times-Union* had not always had that name. When the automobile was in its infancy in the 1890s, the dominant newspaper of Jacksonville was the *Florida Times-Union & Citizen*. Because I have relied so heavily on the *Florida Times-Union* microfilm located at the downtown public library as a source for my books, I became curious as to the history of the newspaper's name. I went to the Jacksonville Historical Society and found a book—Pleasant Daniel Gold's *The History of Duval County, Florida*—that had a brief genealogy, if you will, of this particular daily newspaper.

Without going into too much detail, I believe the derivation of the name goes something like this. According to Gold, the first newspaper in our area came out before the War Between the States. It was called the *Jacksonville Courier*, but it failed, as did many others that followed. After the war, still more newspapers came and failed, but the *Florida Union*, a newspaper with a very Republican slant, was one that survived. It launched near the end of the war, and in 1881, it was purchased by E.M. Chaney, who was the chairman of the State Republican Executive Committee.

Charles H. Jones was a Confederate veteran who had served as editor of the *Electric Magazine* and *Appleton's Journal* in New York, and since the war, he had established and run the *Daily Times* in Jacksonville. He became associated with the firm of Varnum, Jones and Company, and in 1883, the firm bought the *Florida Union* and combined it with the *Daily Times* to get the *Florida Times-Union*. Then, in 1897, the *Florida Times-Union* joined forces with the *Daily Florida Citizen*, and the new newspaper was called the *Florida*

Times-Union & Citizen. (Sometimes the word *Florida* was not used, just as we sometimes drop the word now.)

It took me quite a number of visits to the library, but I was finally able to find the date that the *Florida Times-Union & Citizen* officially became the *Florida Times-Union*. It was January 20, 1903. On the page with the masthead, there was an explanation for the name change:

> *THE SAME IN SPIRIT, BUT A CHANGE IN NAME*
> *In the mottoes now hung in the editorial rooms of the leading conservative newspaper of the country or stored away in the files for the guidance of the staff, may be found the words: BREVITY—CLEANNESS—ACCURACY—AND TRUTH.*
>
> *The three latter the Times-Union will always have as its cardinal law, the first has always been another rule, in all except the name of the print page, which is this morning taken away after being there for five and a half years.*
>
> *Everybody knows that the Times-Union means the Citizen and that the Citizen means the Times-Union, therefore the Citizen is superfluous, and has been removed.*

The explanation continues, noting that because circulation was increasing, the newsboys' lungs were basically taking a beating with having so many words to say to hawk the newspapers. It was "merely a matter of convenience and business, besides being in keeping with some startling improvements decided upon and in sympathy with the laconic spirit of the age." So, there you have it. The *Florida Union* became the *Florida Times-Union*, which became the *Florida Times-Union & Citizen* and, eventually, went back to the *Florida Times-Union*. And so it has remained for well over one hundred years.

JACKSONVILLE'S EARLY AUTOMOTIVE HISTORY

In order to properly cover automotive history in Jacksonville, it is necessary to go back to the beginning. To do that, I consulted one of the ultimate sources on Jacksonville history, T. Frederick Davis's *History of Jacksonville, Florida, and Vicinity, 1513 to 1924*. In it, Davis referred to a "venerable colored man," Sam Reed, who owned a mule named John. Davis said that Reed operated the first wheeled vehicle in Jacksonville. Sam Reed and John "not only did the draying for the town, but it was also the town hearse in the early 1850s. Rowboats supplied the place of carriages; otherwise the people rode horseback or walked. The rowboat came into its own for marooning parties and picnics under the trees on the banks of the beautiful St. Johns River."

Davis also noted that the omnibus (a multi-passenger carriage) and street hack (a carriage for hire, rather like a taxi but with a horse) would make their appearance soon after the War Between the States, and it wasn't long after that the roads of Jacksonville were filled with wagons, drays, buggies and carriages.

Then, on April 30, 1896, a headline appeared on the third page of the *Florida Times-Union & Citizen*, announcing, "The First Horseless Carriage in Florida Received Yesterday." The short article that followed simply stated that this horseless carriage belonged to John Einig of the S.B. Hubbard Company, a hardware and appliance company located in Jacksonville.

Not only was John Einig a shipbuilder, a businessman and the proud owner of the first automobile in Jacksonville, he was also the creator of "Big Jim," the thirty-two-inch steam-powered whistle that even today sounds off

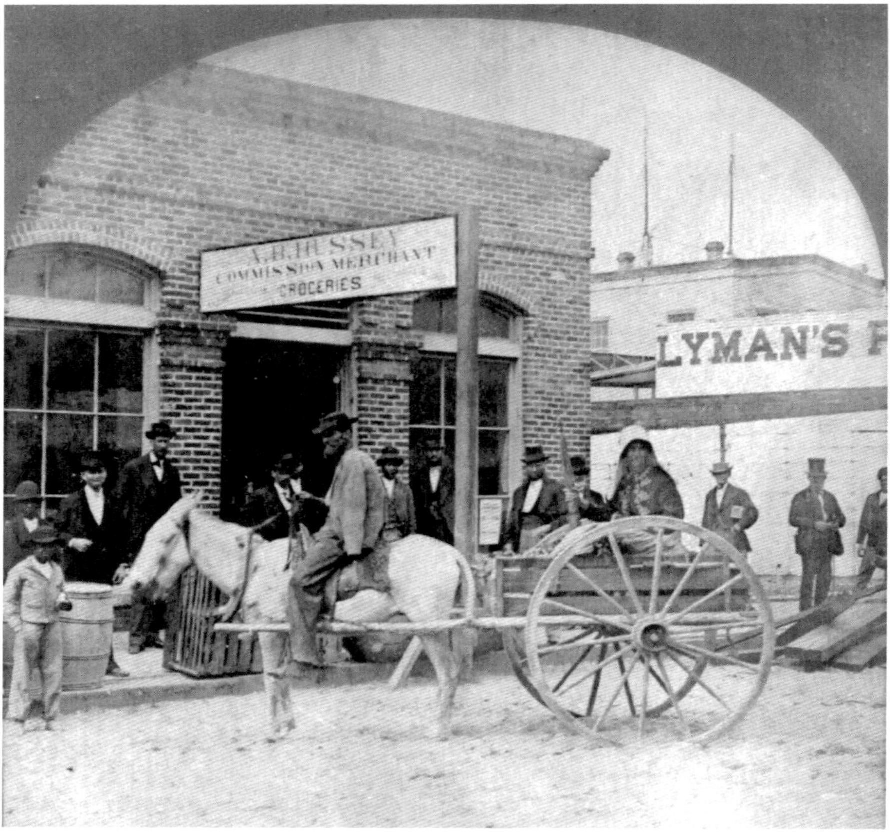

Sam Reed and his mule, John. *Courtesy of Jacksonville Historical Society Collection.*

at 7:00 a.m., noon, 1:00 p.m. and 5:00 p.m. He designed and built this loud steam contraption in 1890 and named it for his brother-in-law, Jim Patterson.

This whistle was originally located at the waterworks until it was moved to the Jacksonville Electric Authority's Southside generating plant. When the generating plant was moved, Big Jim went back to Water Works Park at First and Main Streets. The whistle, which is no longer steam powered, has become a steadfast part of Jacksonville's history and daily life. Even as I was doing research at the Jacksonville's Main Library, I could hear the loud whistle sound off to remind me that it was time to feed the parking meter. On open-window days, a person can still hear Big Jim far into the suburbs of the city. There was even a news story on WJXT showing the Jaguars practicing in the late afternoon sun when the loud call of Big Jim sounded, almost drowning out the reporter.

With only a few interruptions since 1890, Big Jim has not missed a blast. In an obituary provided by the website Find-a-Grave, John Einig's whistle is said to have "sounded the alarm for the Great Fire of 1901, for the end of World War I and II, and for the death of John Einig."

I was able to communicate with Carol Einig Howell, the granddaughter of John Einig who now lives in Georgia. She told me, "My daddy was John's youngest son. My daddy showed me clippings about the family, the car and the whistle. My grandfather also owned a boat called the *Edith*. My daddy talked very fondly of his father. The house that they lived in was torn down to build the Mathews Bridge. I would loved to have seen it. My sister and I were in Jacksonville in August and we went to see and to hear Big Jim blow. It is amazing that it is still in use."

With all that said, this was the very same John Einig who is important to the automotive history of Jacksonville and who was featured in the newspaper in 1896 for his incredible vehicular machine. Einig said that the proper name for his vehicle was "Motor Wagon"; the article described his carriage as looking "very much like a buggy when rigged up with a body [horse]. The boiler and engine are directly beneath the seats and a small water tank is on top of the boiler. The motor wagon is a narrow gauge, but Mr. Einig will lengthen the axels and make it standard gauge. The steering

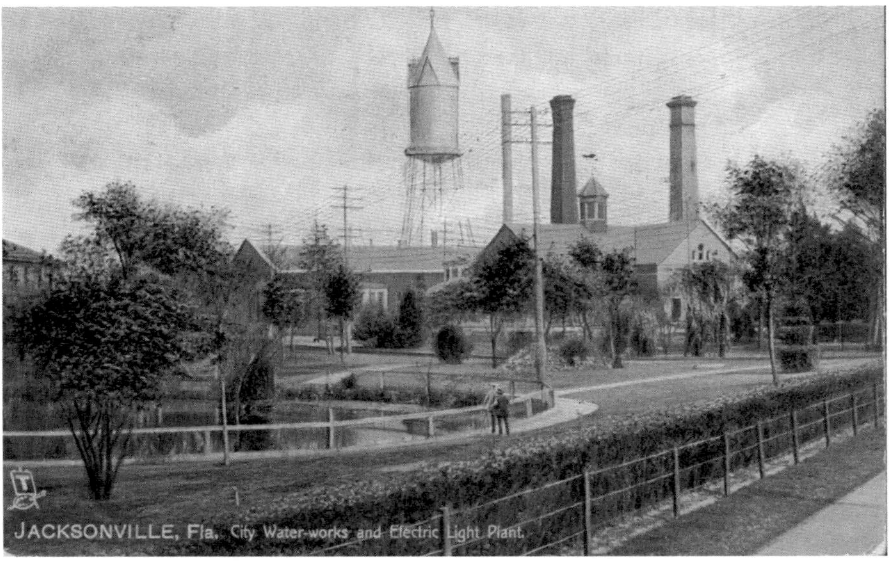

Water Works and Electric of Jacksonville, 1900–1910. *Courtesy of Jacksonville Public Library.*

apparatus is very simple and anyone can soon learn how to manage the entire machine."

This new invention could be fueled with coal, gasoline or naphtha and could be "run at very little expense." Most importantly, the article noted, "When it is first seen on the streets it will cause a sensation." The horseless carriage had yet to be completed, but the newspaper was already certain that it was going to cause a scene. That line alone indicated that Jacksonville was more than ready to become a part of the American automobile revolution.

In Davis's history, he noted that Einig's "autocar" had a steam engine that "threw out a blinding cloud of steam when running and made a noise that caused it to become known as 'Einig's chug-chug wagon.' Its mechanism confined it to paved streets. The heat generated by the engine was so great that it was uncomfortable for long runs."

Einig sold his machine to an Englishman for $1,000, and its subsequent history is unknown. Davis noted that "[a]n illustrated description of this motor buggy was published in the *Scientific American* and the publicity brought many inquiries to the inventor."

John Einig was again in the Jacksonville newspaper on July 2, 1899. The headline in the *Florida Times-Union & Citizen* read, "In a Horseless Carriage—John Einig Will Have an Automobile Ride on the Fourth." He was going to celebrate the Fourth of July by riding through the streets of Jacksonville. This different carriage had arrived in sections on the Clyde steamer *Comache*, and Mr. Einig had it set up in a vacant store on Bay Street, "where skilled workmen under his direction were put to work on it to get it ready in time to take out upon the road Tuesday."

The article goes on to describe the component parts and assembly and notes that the automobile would "attain a speed of eight or twenty miles an hour," an odd fact that seems to show that measuring speeds was still more of an estimate. At the end, the article concluded, "Mr. Einig is much pleased with the carriage, and after making a trial trip, will experiment with it, in order to try and make improvement on the original." Davis, however, said in his history that this "auto-carriage" of French design "did not prove entirely satisfactory and was eventually discarded."

No matter whether John Einig is remembered for his faithful whistle or the earliest horseless carriages in the city, his name is inextricably connected to forward-thinking people in Jacksonville. He was making sensational news with his automobiles in Jacksonville and in the world with his horseless carriage design; of course, we all can be reminded of him four times a day when we hear the blasts of "Big Jim."

CHARLES A. CLARK IS another name that is linked to Jacksonville's early automotive history. Clark was, according to the *Jacksonville Architectural Heritage* by Wayne Wood, a funeral director and embalmer. The Clark Funeral Home was his place of business, and its upper floors were his residence. The building, designed by Thomas E. White, was located at 38 West Forsyth Street, and in later years, it would become the Republic Theatre and then the Roxy Theatre.

In automotive history, Clark is credited with owning the first factory-made automobile in this area. On January 5, 1900, the *Florida Times-Union & Citizen* ran an article with a headline that read, "FIRST IN FLORIDA/C.A. CLARK IMPORTS A MODERN HORSELESS CARRIAGE/ LOCOMOBILE ARRIVES ON HIS ORDER." The article went on to say that it was the "object of much interest and comment" when it "made its appearance in this State."

The machine was described as

> *a rubber-tired driving buggy in its outward appearance, except that there is no allowance made for attaching a horse. It is the size of an ordinary buggy and it is equipped with bicycle pneumatic tires, handsomely upholstered and finished, and it is claimed is capable of running forty-two miles an hour. The machinery, which is of the simplest kind, is hidden entirely. And there is no clumsy appearance that many associate with the machine.*
>
> *Gasoline is the motive power and this is used to drive a 5 horse power engine that is hidden under the seat. The speed is regulated by a lever at the side of the seat, and it is steered with one handle in the middle of the machine. The locomobile is supplied with water and gasoline tanks and can run forty miles on one charge of water and seventy-five miles with one of gasoline. It is practically noiseless, and it is said will not frighten horses. A brake is attached to the rear axle that will stop the machine in a much shorter time than a horse can be stopped.*
>
> *The machine weighs 450 pounds without water or gasoline and 600 pounds when in working order. It cost at the factory $650, and together with the freight and several latest attachments Mr. Clark has supplied, it cost in the neighborhood of $800, delivered in this city.*

It isn't until the last paragraph, however, that the story reveals how much of a stir this contraption was going to cause:

> *Mr. Clark had Chief Haney, Dr. Neal Mitchell, Dr. Hollace Mitchell, Dr. Perry, a Times-Union and Citizen reporter, and several other citizens out in*

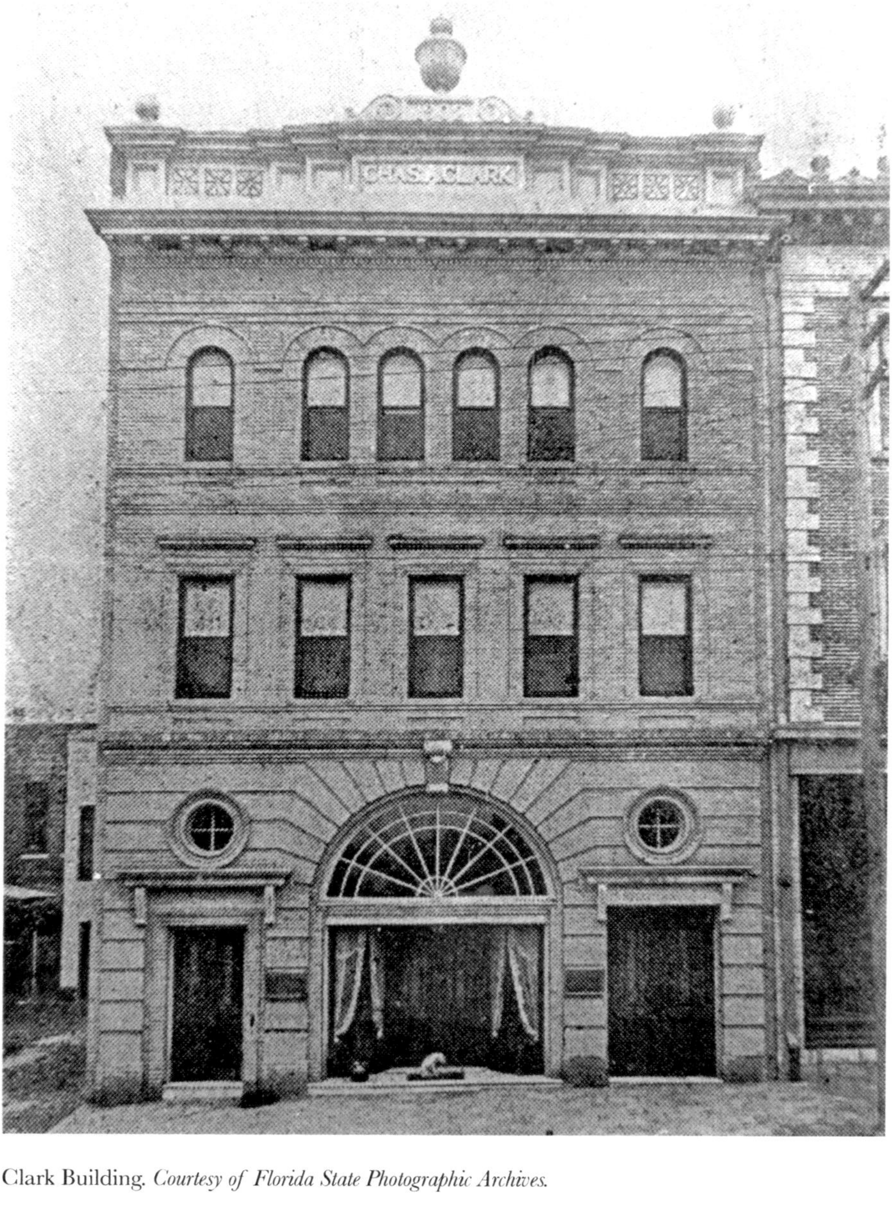

Clark Building. *Courtesy of Florida State Photographic Archives.*

> *this machine yesterday, and all were delighted with the speed and simplicity of the invention. Several professional men announced their intentions of duplicating the order immediately. Mr. Clark will use his locomobile for both business and pleasure, and has assurances of several prominent people here that his lead will be followed by them as soon as the factory is able to supply the machines.*

Even though C.A. Clark's pride and joy really wasn't the first horseless carriage in the area, his seems to have overcome the steam and sound problems that Einig's car had. Yet there is no denying that Clark contributed to car fever that was about to infect the Jacksonville area. The rest, they say, is history. The horseless carriage was here to stay.

On August 12, 1900, the *Times-Union & Citizen* ran a very short article noting, "Swift and Co. have secured an elaborate automobile delivery wagon that was the new thing on the street yesterday." So, it would seem that the commercial use of a horseless carriage was now a reality and was gaining acceptance.

On February 2, 1901, an article ran in the *Times-Union* that noted that five freight cars loaded with automobiles had passed through city en route to Miami, consigned to the Automobile Street and Transfer Company. How amazing that cars on a train merely passing through the city would be important enough to warrant an article in the newspaper! Interest in cars was running high, it would seem. But then tragedy struck.

As everyone in this city must know, one of the worst things that ever happened to Jacksonville was the Great Fire of 1901. Just a year or so after Clark was showing off with his locomobile, Jacksonville was destroyed by fire on May 3, 1901. On May 4, 1901, the *Florida Times-Union & Citizen* announced, "JACKSONVILLE DEVASTATED BY A MOST DESTRUCTIVE CONFLAGRATION." According to the book *The Great Fire of 1901* by Bill Foley and Wayne Wood, seven people lost their lives, 146 city blocks were affected and more than two thousand buildings were destroyed, leaving ten thousand homeless.

With this tragedy causing such widespread destruction, almost everyone here was in some way affected. The *Florida Times-Union & Citizen* encouraged citizens whose homes were still standing to take in those who had nowhere to go. Even then, there wasn't enough lodging for all the displaced persons. Soon, all efforts turned to recovery and rebuilding. Leisure activities or newfangled contraptions were hardly a priority.

Just two years or so after the fire, however, Jacksonville was ready to celebrate. The city held the annual Gala Week, which had been established in 1897, and this time it was quite the event. There was much to celebrate. Foley and Wood noted that the population was growing, as were businesses, and more buildings had been rebuilt than had been destroyed by the fire.

The October 16, 1903 *Florida Times-Union* article titled "AUTO PARADE TO BE A FEATURE/…A PARADE IS DECIDED ON/Owners of Horseless Carriages Will Join Procession—Committee Is Appointed" noted:

> *A grand automobile parade is to be another of the innumerable features of the Jacksonville Trade Carnival.*
>
> *In order to perfect arrangements for the parade a conference was held yesterday afternoon by members of the Florida Automobile Association and the officers of the Carnival Association. A large number of automobilists were present and they expressed themselves as being willing to take part in the proposed parade and to help arrange for it.*

The article also said that it was estimated that there were forty automobiles in Jacksonville, and "with this number, quite an imposing showing could be made….It will be the first time that all of the automobiles in Jacksonville have ever been in a procession, and for that reason, the parade will be especially interesting for local people as well as to visitors."

The parade occurred on November 5, 1903, and the next day, a very short article ran in the *Florida Times-Union*:

> *YESTERDAY'S PARADE—Florida's first automobile parade was a great success yesterday. Thirty automobiles appeared in the parade, which was led by the chairman of the automobile committee, Mr. W.W. Frazier, Mrs. C.W. Purkhul* [this name is an approximation since the microfilm was hard to read], *Mrs. W.F. Spangle, daughter of Mr. Frazier, from North Carolina, and Mrs. W.F. Frazier, all in handsome toilets and among the numerous automobiles in the procession were prominent young ladies of the city charmingly attired in handsome costumes, children in gala dress, and gentlemen of prominence.*

It would seem that Jacksonville was ready to celebrate its recovery from one of the worst of calamities, and the parade was a tour de force. It was almost opulent. All manner of vehicles were represented—horse-drawn, man-powered or horseless—and each entrant's vehicle was strewn with

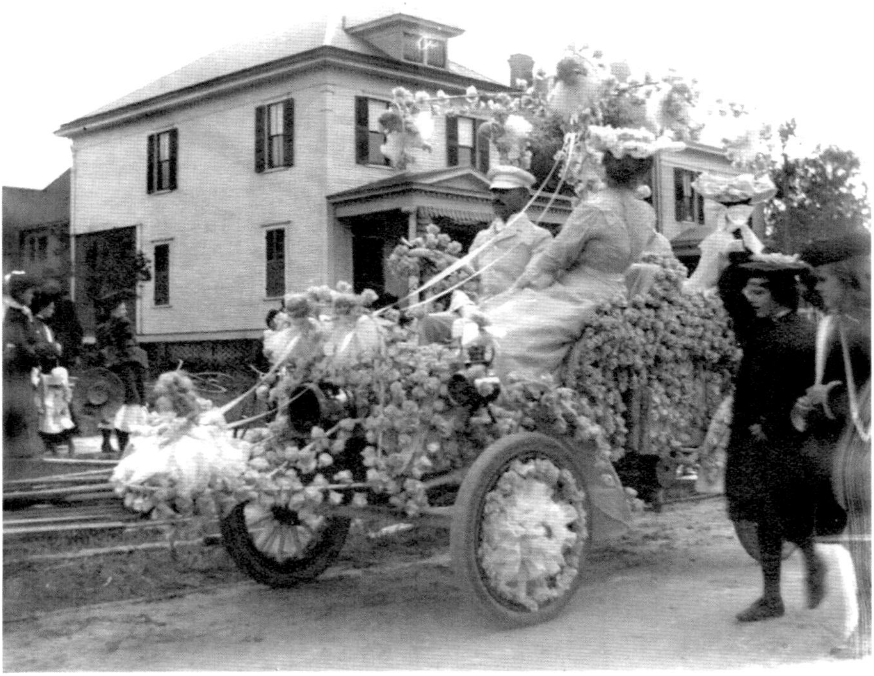

Gala Celebration, 1903. *Courtesy of Jacksonville Historical Society Collection.*

flowers. If the pictures taken by Leah Mary Cox and displayed in *The Great Fire of 1901* are any indication, mountains of flowers were used to decorate all participants. This Floral Parade made its way down Bay Street, and more than six thousand spectators watched. There were thirty-two automobiles represented—the largest procession of cars ever in Florida. Although not all of these automobiles belonged to Jacksonville residents, their inclusion indicated a continued growing interest in these machines.

As noted in *History of Jacksonville*, this was the first such event for cars in Florida. Davis had the parade taking place on November 3, 1903. There were "32 automobiles representing 26 different types and styles in line." (The discrepancy in dates is possibly due to the run date of the newspaper. As to the discrepancy of the number of vehicles, it would be hard to say.)

Still it must have been quite the event to be mentioned in the newspaper and Jacksonville history books. One can only imagine how excited the public must have been to see such an array of vehicles.

AT THE TIME OF the Great Fire, there were only three horseless carriages in Jacksonville. On November 4, 1905, an article in the *Florida Times-Union* declared in its headline that there were "MANY AUTOMOBILES IN JACKSONVILLE/MACHINES REGISTERED WITH RECORDER NUMBER 166." The reporter had apparently gone to the record office and actually listed every car and its maker. There were many easily recognizable companies listed in the article—like Oldsmobile, Cadillac, Ford or Packard—but there were also many more others with less well-known names: Locomobile, Elmore, Century, Orient, Northern, Pope-Hartford, Hayes Apperson, Queen Maxwell, Pope Averly and many more. There was even one "homemade machine" listed. The reporter continued by saying that 151 of Jacksonville's cars were powered by gasoline, 7 with steam and 8 by electricity.

Probably the most interesting fact of the whole article was that a new machine that was yet to be registered belonged to Mr. W.E. Sebree of New York City. He had arrived a few days before with a "large observation car, which will operate in this city for the pleasure of the tourists. It may be added that this is the largest machine which ever came to Jacksonville, having a seating capacity for forty people." The report finally concluded that the "use of so many automobiles in Jacksonville goes to show that the horse is becoming a thing of the past with the business man as well as their families in this city."

With the advent of the automobile, a few unintended consequences occurred that are worth mentioning. One such unimaginable incident happened to Dexter Kelly and was chronicled in the *Florida Times-Union* on June 22, 1910:

> *Now that the vitrified brick paved road to Pablo Beach is practically completed and several automobiles are daily driven to the crescent-like beach, all drivers are warned to keep away from any point near the jetty, as the beach there is treacherous and if the drivers are not very careful they will be swallowed up in the breakers like the two machines last Sunday.*
>
> *Mr. Dexter Kelly, the well known young automobile man, who owned the big Cadillac that got caught Sunday, stated yesterday that at the point where his machine was submerged, is near the south jetty, where there is a kind of slough. It was low tide, and just about the time the machine got in this slough, a monster wave came over the hard sand and caught the machine, holding the wheels as if in a vice. Another wave followed and pretty soon the entire machine was submerged.*

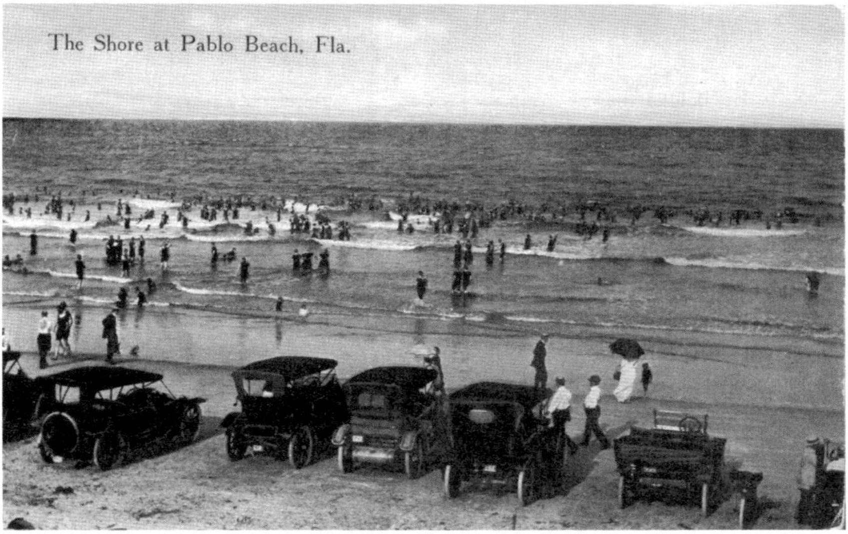

The shore at Pablo Beach, 1900–14. *Courtesy of Jacksonville Public Library.*

The article gave Kelly's warning to all drivers about the south jetty, and it was also revealed that he was not the one driving the car at the time. Thankfully, the machine was pulled out and was "thoroughly overhauled."

Another problem drivers did not foresee was the advent of "joy riding," where thieves would drive off in an automobile just for the joy of driving it (the thieves were undoubtedly unable to afford their own cars). On May 30, 1910, the *Florida Times-Union* announced that Mr. A.E. Baker's Hupmobile had been "brazenly taken" from the front of the Seminole Club the Saturday evening prior.

Even though the article is very hard to read, with the microfilm exceedingly dark and the print very blurred, it was clear to see that owning a car was not going to be easy. Even if it ran well and you didn't leave it in the incoming tide, it was possible that it might be stolen and driven all over Jacksonville until it failed, ran out of fuel or was crashed.

And so it goes, even into these modern times, I suppose. Still, these associated problems were not going to stem the rise of automobile-mania. It hadn't been fifteen years since the advent of Einig's uncomfortable and loud "chug chug machine" in 1896 and already the automobile had become a fixture in the Jacksonville cultural landscape. The handwriting was on the wall—the days of horses and buggies were indeed numbered, and things were certainly beginning to change from here.

ROADS, FERRIES, BRIDGES AND TOLLS

Even though there always seems to be road construction going on somewhere in Jacksonville at any given time, most citizens who drive here seem to take it for granted that there will be relatively sound roads for them to travel. People here hop in their cars and head out to work or the store and simply assume that they can arrive safely wherever it is they want to go. This was not always the case in earlier times in Jacksonville.

According to Davis's *History of Jacksonville*, "The first paving project in Jacksonville and Duval County was in the 1850s when a company of enterprising citizens was formed to construct a plank road from Jacksonville to Alligator Town (Lake City), following the route of the western branch of old Kings Road. Eight miles of plank road was completed, when the railroad (Florida, Atlantic & Gulf Central) became assured, the plank road construction was abandoned leaving the stockholders of the enterprise responsible for debts that resulted in numerous lawsuits."

This road started at Bay Street and Newnan Street and went north on Newnan to Duval Street, west on Duval to Laura Street, north on Laura to Church Street and then west to a farm owned by Isaiah D. Hart called "Cracker Swamp." Davis reported that remnants of this road were still in existence after the War Between the States.

Davis went on to say that in 1882, an editorial was printed in the *Jacksonville Daily Times* that described Bay Street as "a chain of lakes and mud holes." With that in mind, it wasn't long before the Board of Trade, formed in 1884, began approaching the Jacksonville City Council about providing better roads. It wasn't until 1886 that work would begin on this road improvement project. There was much divisive discussion about how this project would be constructed. The council finally picked Wyckoff cypress blocks placed on one-inch planks, but not everyone was happy with this idea. Even as workers went about their business, naysayers came to taunt and heckle the workmen. One interesting problem occurred when the FR&N Railway, the streetcar railway, refused to pay its share at the last moment. As a result, the strip in front of its property had no new road.

All went well with this road for a number of years, but then problems began to arise with it. First, it was impossible to be kept swept clean, and the road was also susceptible to fire. In 1891, a stretch of Main Street burned, leaving only the rims of the blocks that made up the road. By 1892, many of the surfaces were becoming uneven since the blocks were beginning to deteriorate outward from the center.

Bricks were to be the answer. In 1893, an ordinance was passed by the city council saying that a new road was to be laid, and this time, the roads were to be made of "vitrified brick on a foundation of shell and cement." Businesses were to pay for the roads in front of their establishments at the price of $2.41 per square yard of paved road. As a result of this ordinance, there was, according to Davis, much "confusion and complaint."

Of course, before the new road could be laid, the old wood had to be removed. This wood was of considerable concern since people feared that it harbored contagion. As a safeguard against illness, the wood was coated in lime and then removed and sent to the "cremator" for destruction.

On June 1, 1894, Bay Street was opened from Bridge (Broad) to Market and was the "finest paved street in Florida at the time." By the end of 1896, 6.5 miles of city roads were made with vitrified brick, 6.7 miles were made of marl and rock and 3.5 miles were made with shell.

One interesting aspect that Davis discussed was his observation that when it came to taxpayer dollars, citizens in the past (as well as citizens today) were not always in agreement as to how these dollars need to be spent:

> *The subsequent history of street paving is written in the terms of bond issues and more miles of paving—and politics. The system of apportioning the paving schedules among the various wards not only produced in the Board of Public Works and the Council, but contributed immensely to defeating any general, well-laid plan to develop thoroughfares. The matter of selecting streets to be paved caused more wrangling in the past than all other public improvements combined. The development of well-defined*

> *arterial roads and streets is a matter of comparatively recent years. In 1924* [the date of the history's publication], *the city of Jacksonville had 350 miles of streets laid out, of which 106 miles were improved.*

Of course, there are so many roads in the North Florida area that another entire book could be written on the subject of their development. Instead, I have decided to write about only one more. It is the road for which I am most grateful, the road that the *Florida Times-Union* called "Florida's Famous Road to the Sea." Atlantic Boulevard was our first paved road to the ocean. Many of us who love the beach and all that it has to offer in the way of recreation are certainly grateful that forward-thinking people made it possible for us to get there safely.

When the road was opened on July 29, 1910, two articles ran in the *Florida Times-Union* about it. One was about how Atlantic Boulevard came about. Some eighteen years earlier, in 1892, Mr. Eugene F. Gilbert, with the help of four other individuals, purchased six miles of oceanfront south of Pablo Beach, which was then known as Burnside. They intended to build a shell road from the Jacksonville ferry to the Atlantic: "This project, however, was abandoned on account of the death of two of the promoters and as the years came and went, all the survivors with the exception of Mr. Gilbert and Dr. Davis, disposed their holdings. These two, however, were bent upon the establishment…of a hard road to the beach."

After about two years of effort and after much money had been spent, work on the beach road was halted. It is hard to tell why exactly, as the article merely says that "the Grim Reaper had not been idle." (It is not at all clear which important persons had died. The article intimated that it might have been county commissioners since the personnel of the commission had undergone a big change.) Some of the new commissioners wanted the project to continue, and others did not. Eventually, the work was abandoned. It would take quite a number of years before the project was again brought back to life:

> *Eight years ago* [1902], *Fred E. Gilbert, the pioneer automobile dealer of Jacksonville, conceived the idea of making the great wide stretch of beach lying but a few miles away a race course, and with the determination that is akin to accomplishment set out to finish the work left off by his father. He knew it was a stupendous task—but he knew no such word as fail.*
>
> *Many hours—days—were spent in surveying the dense swamps that stretched out between the ferry and the beach. But the road was still a*

> *possibility—and it was to be made a reality. The attention of the county commissioners were directed to what might, with the proper superintendence be acquired. The aid and influence of the board of trade was soon enlisted and with Herbert B. Race, and Charles A. Clark* [first owner of factory-assembled automobile], *the interest was so thoroughly aroused that a committee of three free-holders, Messrs. W.W. Cummer, Charles E. Garner and W.M. Archibald, were selected to go over the road and select what they thought to be the most desirable route.*

From then on, the project continued, even though there was much to be done. Fred E. Gilbert was able to procure convict labor to do the work. He was also able to get a bridge built at Pablo Creek, a difficult feat that required "the best of engineering—there was nothing but dense swamp to both approaches."

Slowly, ever so slowly, the road inched its way toward completion. Sadly, Gilbert's father, Eugene, died on February 8, 1907, about two and a half years before his dream of a beach road could be realized. As the *History of Jacksonville* noted, Gilbert was an automobile "enthusiast without peer. Atlantic Boulevard to the beach was largely the result of his enthusiasm and persistent effort."

On July 29, 1910, Atlantic Boulevard was opened to the public. Through the efforts of two organizations, a parade was planned, and the portraits of these organizers were included with the articles in the *Florida Times-Union*. There were five Duval County commissioners: H.A. Hudnall, C.F. Flynn, Chairman F.J. Hyde, W.J. Picket, D.J. Conroy. There were four officials of the Jacksonville Automobile Club: President W.M. Stinson, Vice-President H.C. Hare, Secretary C.A. Barnes and Treasurer O. Pierre Havens.

There was much fanfare and pageantry that accompanied this event. The banner headline in the second section of the *Florida Times-Union* read, "ATLANTIC BOULEVARD CELEBRATION." The following headline briefs read, "AUTOMOBILE PAGEANT MOVED TO THE STRAINS OF MARTIAL MUSIC/MILITANT AND SPECTACULAR PROCESSION MARKED FORMAL OPENING OF ATLANTIC BOULEVARD YESTERDAY. MANY GAILY BEDECKED AND BERIBBONED CARS MOVED THROUGH STREETS."

The article went on to say that there were some fifty automobiles in the procession, many of which were decorated with flags, ribbons and flowers. The automobiles were assembled initially along Monroe, east of Hemming Park, in a procession that stretched for some five or six blocks. The parade was led by a police escort: two "big touring cars," one red and one black, and each bore a number (854 and 797). There is no mention of which number went with which color. The cars were driven by a police squad that the newspaper called "the pick of the department."

The parade was not without drama or incident, however. A two-horse truck collided with the band wagon at Main and Forsyth, "the heavy tongue sweeping the rear of the auto truck, upsetting many of the musicians and throwing Groover Girvin to the pavement breaking his right leg at the ankle."

Still, the parade continued down to the ferry, where each automobile was ferried across the St. Johns River to the south side. It took until 3:00 p.m. for the last parade automobile to be ferried across the river. (I was unable to ascertain how many automobiles that a ferry could safely carry at one time, but it seems that it wasn't very many. In Davis's history, he noted, "Before the Jacksonville–St. Johns River Bridge was built in 1921, it was not unusual on Sundays and special occasions, though two large ferryboats were in use, for a line of automobiles awaiting their turn to cross the river to form reaching from Broad Street to the ferry slip at Main.") Once the cars arrived on the south side of the river, they continued to Little Pottsburg Creek. There Miss Marie Hyde, the sponsor of the parade, christened the Atlantic Boulevard.

Finally, everyone made it to the terminus of the boulevard at the ocean to participate in more festivities and exercises. The *Florida Times-Union* noted that Atlantic Boulevard "at last is really a reality and a joy to all good road and motor enthusiasts."

One can only imagine what a journey by automobile on this new road would have felt like. At the tail end of July, I am certain it was hot and muggy and fraught with insects and dirt, but it had to be quite exciting to be part of automotive history. My hope is that everyone stayed the night at Pablo Beach and made the long trek home the next day when they were sufficiently rested.

One of the more interesting things I learned as I researched the construction of Atlantic Boulevard was that for many years, Jacksonville depended greatly on ferryboats to get its citizens over the river.

Several ferryboats are mentioned by name in Davis's *History of Jacksonville*. The *Topay* and the *Fanny Fern* worked different landings near Jacksonville in the 1860s and early 1870s. There was the *Louise*, which connected railroad terminals, and there was the *Uncle Sam*, which ran a route between the city and the JStA&H Railway at South Jacksonville.

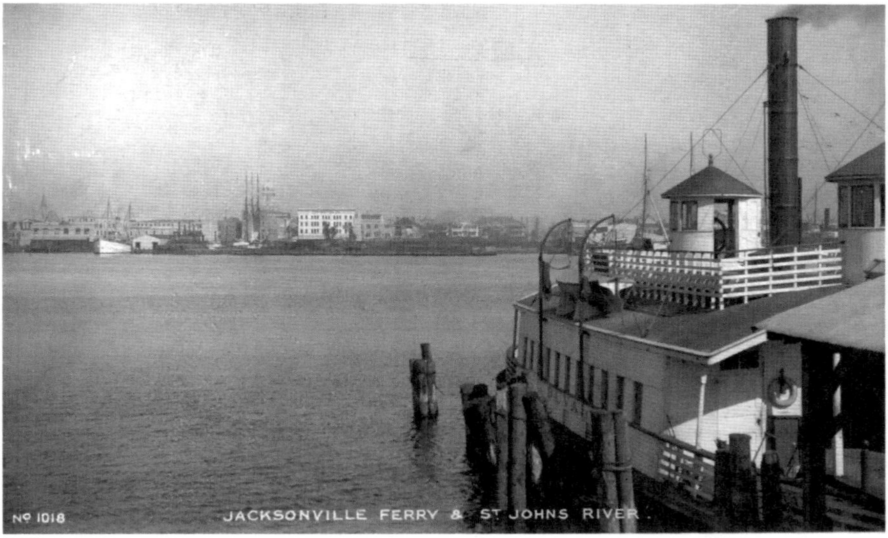

Jacksonville Ferry/Duval Ferry, 1895–1910. *Courtesy of Jacksonville Public Library.*

There were several more ferryboats that moved people and eventually cars from the city to South Jacksonville and back: *Armsmear*, *Mechanic*, *Ravenwood*, *Commodore Barney*, *Duval*, *Dixieland* (a small naphtha-fueled boat) and *South Jacksonville*.

The history goes on to say that "when the waterfront of South Jacksonville was bulkheaded and filled in, the remnants of many of them [ferryboats] were covered up, as that was a favorite dumping ground for those worn out in service."

Thankfully (or maybe not), Jacksonville found it necessary to find a faster, more efficient way to cross the St. Johns than a ferryboat, and Jacksonville's bridges would soon become part of the roads system of the city. The first bridge built over the St. Johns River near Jacksonville opened in 1921. At that time, it was called the Jacksonville/St. Johns River Bridge. On August 18, 1949, an article ran in the *Florida Times-Union* announcing that the bridge was going to be renamed the St. Elmo W. Acosta Bridge, next to the railroad trestle. The article noted that Acosta, "the late city commissioner and legislator, was the so-called 'father' of the span." Governor Fuller Warren and his wife were in attendance at the renaming ceremony, and at one time, this bridge consisted of three lanes, the center lane having tracks for the streetcar that used it.

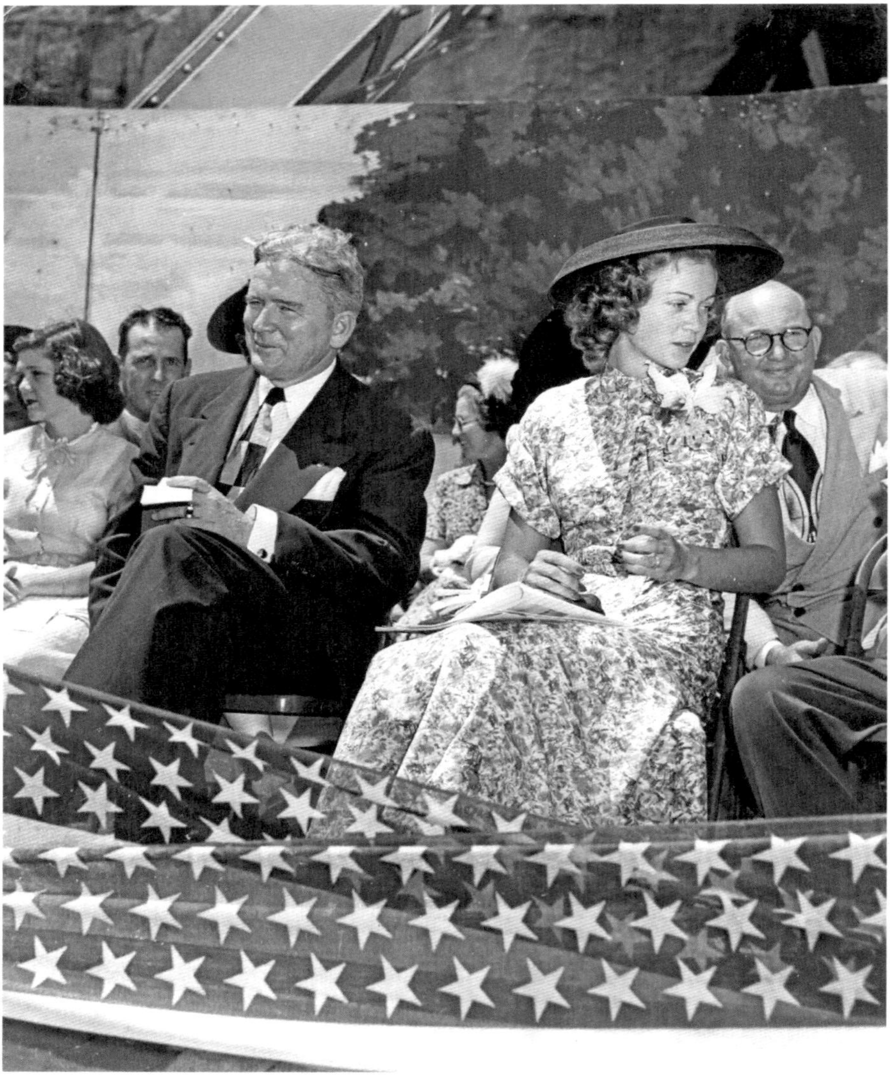

Governor and Mrs. Fuller Warren at the naming of the John T. Alsop Bridge. *Courtesy of Florida State Photographic Archives.*

The next bridge built was what we now call the Main Street Bridge. It is actually named the John T. Alsop Junior Bridge. It was opened in July 1941 and has since become the symbol of our city, even emblazoned on our Super Bowl logo.

The John E. Mathews Bridge was dedicated on January 1, 1953, about one month before it was finally completed. Florida Supreme Court justice John E. Mathews was there to officiate at the ribbon-cutting ceremony, according to an article in the *Florida Times-Union* published on January 2, 1953.

What we now call the Fuller Warren Bridge opened on June 7, 1954. At the time of its opening, it was called the Gilmore Street Bridge. In December of the same year, the bridge was renamed for former governor (and former resident of Jacksonville) Fuller Warren. Many people can recall the first Fuller Warren Bridge for two things: the drawbridge in the center and the tollbooths that stood on the east side of the river. Both things often caused terrible traffic jams! It was with great joy that the tolls were eliminated during Tommy Hazouri's term as mayor in 1988, and it was another happy time when the bridge was redesigned without a drawbridge in 2002.

The Isaiah D. Hart Bridge is another beautiful landmark of Jacksonville, providing spectacular vistas of the city. This bridge was opened on November 2, 1967. The *Florida Times-Union* noted that Jennifer Fewell, who was eight years old and the great-great-great-great-granddaughter of Isaiah D. Hart, "the father of the city," cut the ribbon to open the bridge.

The Henry Holland Buckman Bridge opened on May 1, 1971. An article in the *Florida Times-Union* published on May 2, 1971, noted that Buckman was a former Jacksonville legislator who authored the Buckman Act of 1905, consolidating the state's existing colleges into three universities. He also helped establish the Board of Control for the universities, the forerunner of the Board of Regents. The article went on to say that the "opening of the bridge had been delayed since 1969 by exploding pilings, re-bidding of repair work and unfavorable weather conditions." Apparently, polluted water was reacting with the cardboard-lined center of the pilings, causing them to explode. Thirty-two pilings had been damaged.

Finally, the most dramatic of all Jacksonville bridges has to be the Dames Point Bridge, which opened on September 24, 1988. It towers over the water with its suspension cables, carrying the eye up even higher. It is breathtaking, indeed.

One of the dearest stories connected to any of the Jacksonville bridges has to be the story that ran in the *Florida Times-Union* on September 23, 1988, about Kimba, the seven-year-old Jacksonville Zoo elephant. She was marched across as part of the parade over the Dames Point Bridge at the opening ceremony held at nine o'clock in the morning the day before.

ONE OF THE THINGS I recall from my youth was searching for and collecting as much change as I could any time I needed to go far from home. It was almost always necessary to have money to pay tolls. If I needed to go to Arlington from downtown, I had to cross the Mathews Bridge, and there was a fifteen-cent toll to get there and a fifteen-cent toll to get back. To get to Riverside or to Lane Avenue, where I sometimes worked as a waitress, I had to cross the Fuller Warren Bridge there and back, which required thirty cents. If I wanted to get to the airport and back, it was another thirty cents at the Trout River Bridge. To get to the Gator Bowl or the Coliseum and back over the Hart Bridge, it was thirty cents, and to exit or to enter J. Turner Butler Boulevard near the University of North Florida's campus, it cost ten cents—or twenty cents if you went back the same way.

Of course, all of the small-change tolls went up to $0.25 in the '70s, and that made my contribution $0.50 for a round trip, or $2.50 per week. And that was not the worst of it. The tollbooths often created bottlenecks in the traffic flow, and the cars would back up for miles, it seemed, as we slowly trickled through the booths. Then the smell of car fumes permeated the air when you had to open the window to give the taker the money or to toss the

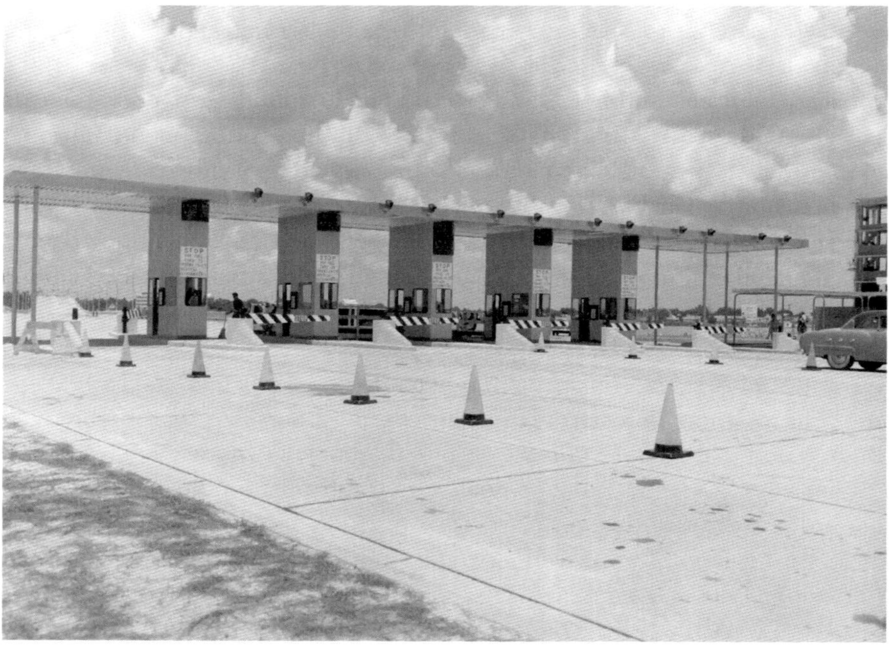

Tollbooths on the Fuller Warren. *Courtesy of Florida State Photographic Archives.*

money into a cone-like contraption that lit up like a Christmas tree if a coin bounced out (which it often did).

It is hard to explain to those who did not have to worry about tolls how incredibly joyous we were when the tolls were lifted. It is reported on the Jacksonville City website (coj.net) that "[i]n March 1989, voters narrowly backed a Hazouri-supported referendum to replace tollbooths with a half-cent sales tax increase; by the end of the year, tolls on the Mathews, Fuller Warren, Hart and I-95 Trout River bridges were a thing of the past."

We have come to take for granted many things when we get in our cars. First, we are almost always confident that they will operate at all. Then we are most certain that the roads we drive on will be navigable. Next, most of us don't have to worry about ferrying our automobiles across bodies of water, except of course for those who use the Mayport Ferry. Most of us use the seven major bridges across the St. Johns River, and many other smaller ones throughout our area, that make traveling great distances easy and safe. Finally, we don't pay tolls to use any of the roads or bridges, and that is a wondrous thing indeed. It is amazing that these things were major concerns in days past, and we can only wonder what marvels we humans will be taking for granted in distant future times. It does beg the question!

A BRIEF HISTORY OF CAR DEALERSHIPS

Few would disagree with the premise that the U.S. automotive industry is the envy of the world. Its volume is the largest and it has a top level of sophistication. Other nations emulate the manufacturing and distribution system in the United States. Yet it is the franchise system that sets the United States apart from other markets. The strength of the automobile industry in this country is 100 years of development of the American auto dealer.
—Keith E. Crain, publisher and editor-in-chief, Automotive News, *September 25, 2006*

When I began researching Jacksonville's car history, I realized that brilliant and driven people (pardon the pun) were the change-makers of our entire culture. Henry Ford, Henry Leland, Ransom Olds, David Dunbar Buick, Walter Chrysler and many others were visionaries who could see the world as a place without horse-drawn transportation—a new world with automobiles that even members of the middle class would be able to afford and drive for business and pleasure. In my generation, Steve Jobs and Bill Gates were the same kind of visionaries who could see a world where everyone had a personal computer. This, in turn, would lead to everyone having a phone that could carry more information than earlier computers that filled entire buildings.

We owe a great debt to these kinds of visionaries for making our lives so much easier and convenient. But make no mistake about it, these people were not only in it for the good of humanity. Fortunes were to be made and were made. Sometimes the path to success is not always pretty or fair. As I read many times in my research, "Business is business." That maxim forgave a host of ills and assuaged a lot of guilt.

In order to understand the evolution of the automobile industry in America generally, and Jacksonville specifically, I consulted a well-documented, special issue of *Automotive News*, published on September 26, 2006. One article, "The Franchise System," written by Leslie J. Allen, chronicled how automobile distribution in America evolved. There was also a timeline that ran along the bottom of the entire issue, which helped me to put into context the development of automobiles and the people who made them.

First, in 1879, George Selden applied for a "master plan" patent for the basic features of an automobile. In 1895, Selden was granted a patent. In 1886, a time somewhere in between the application and the granting of Selden's patent, Ransom Olds built his first horseless carriage and Henry Ford invented his first car, the Quadracycle. In 1897, Olds organized the Olds Motor Vehicle Company, and in 1902, Henry Leland formed the Cadillac Automobile Company. In 1903, Ford incorporated the Ford Motor Company, and David Dunbar Buick created the Buick Motor Company.

Now, here is where things get a bit sticky. In 1903, a group of automakers formed an association that recognized the Selden patent, and they agreed to pay Selden royalties. Henry Ford declined to do so and was then refused membership in the Association of the Licensed Vehicle Manufacturers. As a matter of fact, the organization sued Ford for patent infringement. In 1908, a judge confirmed the legitimacy of the Selden patent, but in 1911, the Supreme Court of the United States overturned the Selden decision and found in Ford's favor. Then, on December 1, 1913, Henry Ford's assembly line produced its first car at the Highland Park Plant in Detroit, Michigan, and the possibility of "a car in every garage" became quite possible.

In the earliest years of the automobile, people like Einig were making their products for themselves and then for their friends. As the industry grew and factories were formed, enterprising and persistent people like Ford innovated a way for many cars to be made at one time. That necessitated a retailing system including "third-party retailers."

Leslie Allen's premise is that "the franchise system grew out of geographic necessity." Automobile makers and their factories were concentrated mainly in places like Detroit or Cleveland. There was a need to get factory-made cars to faraway buyers in places like Jacksonville who were clamoring for automobiles. One of the earliest ways the major companies used to get cars to the public was called the "branch system." This system worked by having company people go with the product to distant areas to run car stores. It eventually proved inadequate, as the automobile market grew more competitive.

The Allen article cited Robert Casey, curator of the Henry Ford Museum, who said that branch managers drew a straight salary and were not paid according to the number of cars sold. As a result, they "weren't as enterprising as the dealers who if they didn't sell any cars didn't make any money."

Eventually, the automobile manufacturers developed a distributorship system wherein distributors became the secondary wholesalers (the factories being the primary wholesalers). An article by John R. Teachen Jr. in the same *Automotive News* noted, "Eventually, the factory took over the whole process. But the distributor set up died hard in some quarters. As late as the 1960s, Cadillac still had a flock of distributors as well as factory branches. Cadillac dealer contracts were labeled 'direct dealer' or 'dealer under distributor.'"

Early in the history of the automobile, some people rushed to be part of this lucrative new juggernaut, even though a great burden was sometimes placed on these dealers. Initially, the dealer investment was small, but he had to buy a demonstrator car, and that was a considerable expense at that time. Customers needed to be able to inspect these machines that they had only seen from afar or had only read about.

Allen cited Charles Mason Hewitt's 1956 book, *Automobile Franchise Agreements*, which delineated the responsibilities of dealers or franchise owners. In the early days, the franchise agreement made auto dealers "exclusive agents" of automobiles for one year at a time. They had to maintain a suitable showroom and repair shop and keep a reasonable inventory. They agreed to sell in a prescribed territory, and they paid 10 percent of the list price up front for each vehicle ordered. Automakers reserved the right to pull a franchise with thirty days' notice if the dealer failed to meet factory expectations.

Jacksonville was certainly ready for the automobile at the turn of the century. Not only did people want to keep up with other thoroughly modern cities in the United States, but they also wanted a better way of transporting themselves and their goods. Horses had proved to be reliable transport for countless years, but they required space and diligent, daily care. They could not always be fast, and they certainly were "messy." Jacksonville wanted and needed modes of private and commercial transportation that were new and different. Because there was desire and need for automobiles, that necessitated the creation of a "broker" or an "agent" to supply them. Agents were the precursor to dealers.

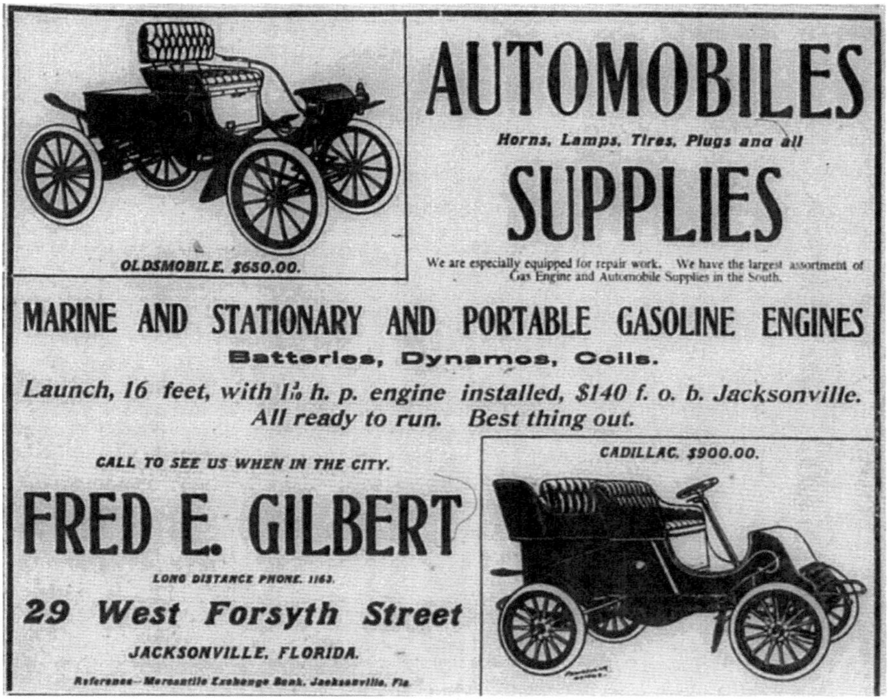

Gilbert advertisement in the *Times-Union*. *Courtesy of* Florida Times-Union.

In Davis's Jacksonville history, there is mention of a Fred E. Gilbert, who opened the first garage (the same Fred Gilbert who was instrumental in getting Atlantic Boulevard built). Davis said that Gilbert was probably the closest thing to an automobile dealer in Florida, and he certainly was the first ever to run a car ad in the *Florida Times-Union*. His advertisement ran on October 25, 1903. Not only did he have all the supplies one would ever need to drive a car, but he also had the cars themselves: Oldsmobiles for $650 and Cadillacs for $900. His place of business was at 29 West Forsyth Street, but the ad didn't have a fancy logo—just his name. Apparently, in those days, that was quite enough.

According to *Jacksonville Revisited* by Andrew Bachmann and Maria Mediavilla, in association with the Jacksonville Historical Society, Gilbert built and then moved his operations to a new, three-story building with showrooms and garages for repairs, storing and washing at the corner of Laura and Church Streets.

There is an earlier article in the *Florida Times-Union* wherein Fred E. Gilbert's name is featured. The article ran a few months earlier on May 20, 1903, and it was headlined, "AUTOISTS FIGHT BILL." A bill had been introduced and passed by the Jacksonville City Council that limited the speed of automobiles within the city limits to six miles per hour. A.G. Hartridge was an attorney for the Florida Automobile Association present at the meeting, and the article noted that his argument was that it would "do the automobile a great injustice, as the machines could not be operated to advantage if they were required to keep speed below six miles an hour. An ordinary man, he stated, could walk very nearly as fast as the time named in the ordinance."

F.E. Gilbert was called as an expert, and he "gave the Councilmen some information as to the mechanical working of the machine and stated that New Jersey, which State was regarded as having the most perfect law on the subject, automobiles were allowed to run a mile in six minutes [ten miles per hour] in the filled up portions of the city, and mile in three minutes [twenty miles per hour] in the less populated portions." It was decided then "to refer the bill back to the committee to make further investigation."

Regardless of his reputation as to the merits of the proposed "speeding car ordinance," Gilbert was what we would refer to as a dealer, and his business was a dealership. From his inauspicious beginnings on West Forsyth Street, Gilbert's product would soon be putt-putting around the roads of Jacksonville, leading soon to more and more dealerships bordering the roads of our city—Riverside Avenue, Main Street, Philips Highway, Cassett Avenue, Blanding Boulevard—and all manner of dealerships in between. Some big and some small, some new cars and some used, but they were always there to offer the innumerable joys of driving.

IN THE 1915 *TELEPHONE Directory for Jacksonville and Pablo Beach*, which is the oldest telephone directory found in the Jacksonville Main Library, Gilbert is listed along with sixty-three other businesspeople under the heading "AUTOMOBILE DEALERS, ACCESSORIES, SUPPLIES, REPAIRS, GARAGES." From Acme Garage at 88 West Beaver Street to Ye Auto Shop at 621 Main Street, these businessmen were ready to supply the automotive needs of a fast-growing, car-loving culture.

In this phonebook, there were businesses listed that sold tires, while others provided upholstery or actually sold the cars. In an additional section entitled "Autos for Hire," there were eleven more listings for businesses that rented out automobiles.

One thing was immediately interesting to me. These businesses had phone numbers that were, at most, only four digits long. Fred Gilbert was then listed at 221 West Adams, and his telephone number was "4106." Occasionally, there would be a listing with only three digits, like the Norton Buick Company at 20 East Orange Street, with the phone number being "347." Claude Nolan Cadillac at 1001 Main Street had a phone number of "24," and in the "Autos for Hire" section, Home Auto Service had the shortest phone number of all: "6."

Another fact that became apparent was that most of these businesses were located downtown. There were automotive businesses on Adams, Bay, Beaver, Cedar, Duval, Forsyth, Hogan, Laura, Main, Monroe, Riverside and Walnut. Many of the listings were simply names—like Fred E. Gilbert's on West Adams or Louis Slawinski on West Seventh. These businessmen must have been very confident to use only their names when promoting their products. It is also possible that some of these individuals' names were listings for residences and that these people made cars in their own yards or garages.

To look at these entries, it is difficult to know which of these addresses were actually selling cars for far-off markets and which were simply providing supplies or repair work. Clearly, Atkinson Tire and Supply Company was not what we would call a car dealership. The same might be said for Acme Garage, Jacksonville Rubber Company and Searchlight & Sunlight Gas Company.

There were, however, others that were definitely selling cars: Aultman Motor Company "Hudson," Jax Auto Clearing House and Maxwell Motor Sales Corp. All of this information could easily fit on three-fourths of a page of the pamphlet-sized phone book.

Jump ahead fifty years to 1965 and there are 130 new car dealers listed in seven columns of telephone numbers in the *Southern Bell Telephone Directory for Jacksonville* of July 1965. Dealers selling new cars filled three pages, and there were numerous advertisements along the sides of each page. All telephone numbers were now seven digits long, and locations had begun to spread away from the core city and down major thoroughfares out into the suburbs.

There are still many listing in the downtown area—at Adams Street, Ashley Street, Eighth Avenue, Forsyth Street, Hogan Street, Haines, Laura Street, Main Street, Monroe Street, Park Avenue, Pearl Street, Riverside Avenue and State Street—but there is now a whole new demographic on the south side of the St. Johns River. There were now dealerships on Beach Boulevard, Hendricks Avenue, Kings Avenue and San Marco Boulevard, but many dealerships were now also located on Philips Highway/U.S. 1 (more than twenty-one listings according to my count, about the same as the dealerships listed on Laura Street and Riverside Avenue combined). Things had begun to change; as the population grew and moved out into the suburbs, so did the number of automobile dealerships.

Today, in 2018, as I write this, the dealership landscape is very much different from the early days of automobile dealers. For one thing, dealerships are no longer listed in traditional phonebooks alone. They are listed online or in abbreviated *YP (Yellow Pages)* books that come to residences as phonebooks used to do. My *YP* has 102 listings for new car dealerships in the Jacksonville area, including in Baldwin, Fort George and Julington. Downtown markets are practically nonexistent, with the People's Car Company being on Duval Street, but with a phone number that only goes to voicemail.

Now, the lion's share of dealerships is located on Atlantic Boulevard, Blanding Boulevard, Cassett Avenue, Philips Highway (although most of these dealerships have moved south of the Avenues Mall) and Southside Boulevard. And so it goes.

Of course, the Internet has totally changed the way things can be done concerning the purchase of a car. Potential buyers can go online to any dealership website and see new models, used inventory, prices and customer reviews. They are able to do a fair amount of shopping before they get directions to the showroom and actually go there. Buyers can feel better informed than they may have felt in times past. And I hardly know what to say about being able to buy a car in one's kimono or out of a vending machine, as Carvana has built. Regardless of how one does it, though, buying a car is still a serious, emotional and exciting thing to do for most people.

After having perused all manner of phonebooks and directories, I can say that an awful lot of industry and effort have gone into replacing the horse as a mode of transportation. As I interviewed many people in the trade, it became even more obvious that hard work is the cornerstone of the dealership paradigm.

DEALERSHIP PIONEERS

At the corner of Atlantic Blvd and Arlington Road was Obi's Garage and Repair Shop. Obi's had about six new Datsun cars in a dirt lot next to his garage, and when I was just 16, I stopped by there several times to look them over.

Mr. Obi came out, and said if I wanted to buy one, he would make me a deal as he was moving down the street on Atlantic Blvd to open a "dealership" with an inside showroom. The following week I rode my bicycle from Ft. Caroline Club Estates up to his new showroom. Over in the corner was a Datsun 510 station wagon.

I told the salesman I wanted to buy the car. He looked at me and said, "Are you serious, young man?"

I took out, from my homemade Boy Scout wallet, 5 crisp 100 dollar bills. He says, "Step in my office."

I told him I had a job, a credit card and a bank account. I asked him to throw in a radio, which he did, and some floor mats. He said I was killing him on the deal. He says he couldn't let me have the car unless an adult cosigned the contract. I asked to borrow his phone and I called my mom. He called the bank and several minutes later my mom drove up. I told her the payments were going to be $45.00 per month for three years, and I could easily make this sacking groceries and cutting lawns. She signed on the bottom line and the salesman congratulated me for buying one of the first Datsuns out of the new dealership. He told my mom he had never sold a car to a 16 year old kid before.

I put my Schwinn bicycle in the back and drove proudly out of the dealership in my brand new Datsun station wagon.

—John C. Christian

If there is one thing to be learned about car dealers in the Jacksonville area, then it is this: they all have very similar stories to tell. From the very earliest days of the automobile, car dealers have worked tirelessly to provide the best their manufacturers had to offer to the citizens of Jacksonville. In good times and bad, these hardworking souls have done whatever was necessary to get people behind the wheel of the finest cars available, all while providing their own families with the necessities of life.

Many of the early dealers came from impoverished situations brought on by the Depression or war. Most dealers here have come up through the ranks, learning all aspects of the family business from washing the cars and preparing them for sale to servicing them, repairing them and selling them. All Jacksonville automobile dealers have worked and do work long hours, and they have built businesses to be admired.

Another admirable quality shared by all dealers with whom I have talked or about whom I have read is that they are very actively involved in the Duval County/Jacksonville/North Florida community. Dealerships sponsor races and teams and give generously to many causes and charities. They are highly visible in this regard, helping make our corner of the world better—first by providing us with automotive transportation and then by supporting all kinds of charitable causes in the area. Dealers are an integral part of this community, and thanks to them, we here in Jacksonville can be part of the American experience.

In the writing of this book, I readily confess that I have left too many memorable dealers out. I have totally left out used car dealers, and then there were the dealerships whose owners were unavailable to be interviewed for any of a variety of reasons. To those I have neglected, I apologize. I had no intention of leaving anyone out, but circumstances have conspired to make omissions happen. For that, I am truly sorry. What I hope to have done is found the pioneers of this business—those whose histories go way back into the last century. These are some of the longest-lasting dealerships in Jacksonville, and as a result, their impact is the most deeply felt.

CLAUDE NOLAN CADILLAC

One of the best things I remember about my husband Bobby's years of selling cars was that he would buy used Eldorado convertibles and then fix them up to sell at a profit for extra family money.

For the entire time he was working on them—taking them to Hathorne's Seat Coverings and to mechanics to tune up the engines—we would use them as our own cars. We would go on great adventures to Fernandina to the Shrimp Festival or to Jacksonville Beach. I remember going through the drive-thrus to get food and people "oohing" and "aahing" about our ride. We would even lend our cars to Wolfson High School when our kids went there, so that the homecoming court could be in a parade.

—*Kathy Owens*

When I was looking in the *1915 Telephone Directory for Jacksonville and Pablo Beach*, there is one, and only one, listing in the automotive section that is still in operation today: "Nolan, Claude, Cadillac Auto." By 1915, Claude Nolan had already been in the automotive business ten years, because it was in 1905 that Claude Nolan became a Cadillac Motor Car agent in the city of Jacksonville. Actually, it is one of the longest-lasting car dealerships in the entire nation.

According to a pamphlet prepared in 2005 to celebrate Claude Nolan Cadillac's one-hundred-year existence in Jacksonville, Nolan was born in Sanford, Florida, and moved to Jacksonville as a child. He graduated from Duval High School, and he earned a law degree from Vanderbilt University. He returned to Jacksonville, where his father, George Nolan, had served as mayor from 1903 until his death in 1906.

Claude Nolan opened his dealership in a small building on East Church Street in 1905, and by 1910, he had moved to a new building designed by

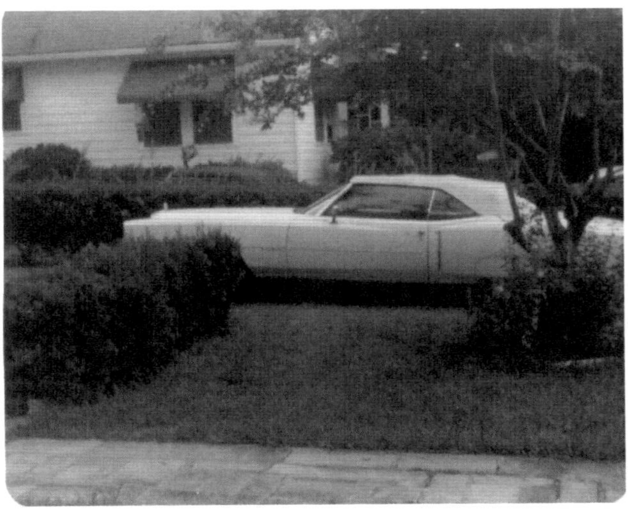

An Eldorado in which Bobby Owens took his family on adventures. *Courtesy of Kathy Owens.*

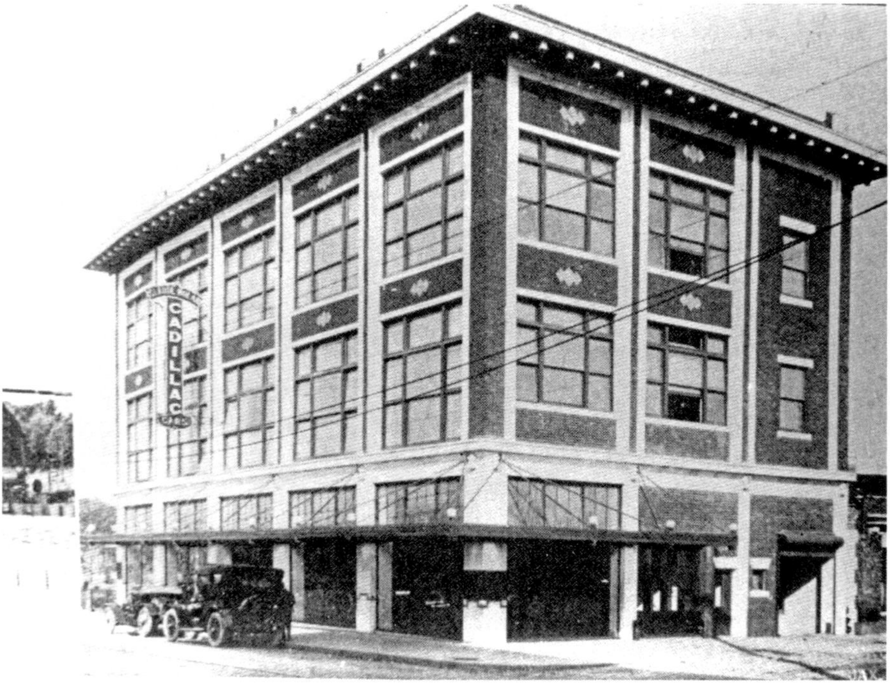

Claude Nolan Cadillac Building. *Courtesy of Florida State Photographic Archives.*

Henry J. Klutho, one of Jacksonville's prominent architects of the time. The three-story dealership at 937 Main Street was quite opulent at the time, with many amenities and a striking mosaic of the Cadillac crest gracing the showroom floor. This building served as the dealership's home until 1985, when it moved to its present Southside Boulevard location.

Cindy Pittman of E.H. Thompson Corporation once worked in the old Claude Nolan building after the dealership moved. According to Pittman, E.H. Thompson Corporation moved in, and the property, then owned by Henry Moats, was renovated by Kent Moats for the new tenants.

When I asked Pittman if the old Cadillac mosaic was left on the floor, she said, "Oh, yes, that mosaic was up the stairs and the renovation did not alter it. The whole thing was very nice. It was a beautiful space to work in."

Sadly, the old building is presently boarded up, but the new Claude Nolan Cadillac location continues the tradition of opulence and beauty in its complex on Southside Boulevard. With huge windows and woodsy grounds allowing light to filter in gently, it is quite elegant.

Of course, during the first forty or so years, Claude Nolan was exceedingly busy. Not only was he selling Cadillacs in Jacksonville, but he had also become a distributor and was able to provide cars for all of the Florida and south Georgia Cadillac dealerships. After World War II, his was the largest single Cadillac distributorship in the nation. He then built a dealership on Biscayne Boulevard in Miami that many called "Nolan's Folly," but he was not deterred. That dealership became "one of the biggest Cadillac dealerships in the nation." He also acquired many Pontiac franchises in Miami and Jacksonville and operated several automotive trade services.

At the time of Nolan's death in 1943, a *Florida Times-Union* article said that he was president and treasurer of Claude Nolan Inc. He had branches in Miami, Tampa and Savannah. He was also a past president of the Cadillac Old Guard, composed of pioneer distributors of the Cadillacs. He was active in promoting good road development and was instrumental in establishing the Dixie Highway, "one of the first good long roads in the South."

In *Jacksonville's Architectural Heritage*, by Wayne W. Wood, there is mention of Claude Nolan as being the president of the San Jose Company in 1914:

> *Nolan was a strong promoter of better roads and modern transportation. The San Jose Company purchased about 1000 acres in this area, but due to the economic downturn caused by the beginning of World War I, the development never got off the ground* [until a decade or so later]. *But Nolan believed that the increasing importance of the automobile would foster the growth of outlying suburban communities.*

One of Claude Nolan's many claims to fame, other than his automotive empire, was mentioned in a *Florida Times-Union* article written about him at the time of his death: "He had the distinction of being the first man to drive an automobile to the City of Key West. The feat was accomplished by driving the car over the railroad trestles connecting the Florida mainland with Key West. That was in 1927. Several years later a hurricane wrecked the trestles and the railroad was abandoned. Later a highway was built over the railroad right-of-way into the Island City."

In the 100th anniversary pamphlet, there is also mention of Nolan racing a Cadillac against an early "aero plane" in 1910 at the Jacksonville Fairgrounds. In the *Jacksonville Family Album*, the picture featured had the following story included:

> *A Day at the Races: Above, May 20, 1910 was a day to remember at the Moncrief Racetrack as it included the second successful flight in Florida. The pilot, Charles K. Hamilton, electrified the crowds with a "death-defying" glide that included cutting the engine. Hamilton also challenged Cadillac Dealer, Claude Nolan to a race above, pitting his bi-plane against a Cadillac driven by Dexter Kelly* [of swamped Cadillac fame mentioned earlier]. *The Cadillac won. Hamilton attributed the loss to a headwind. Nevertheless, the winning owner showed true sportsmanship and even more courage. He accepted a victory lap in the plane, and according to aviation historian, John Ingle, Jr., Nolan became "the first native Floridian to fly over Florida soil in an airplane."*

The pamphlet noted that Nolan never married, but he looked after his widowed sister, Lila Nolan Peterson, and her daughter, his niece, Claudia. Claudia grew up and married Connor Brown in 1938, and Brown joined the business. When Nolan died in 1943, he was succeeded by Brown as distributor. The company continued to operate as Claude Nolan Inc.

The pamphlet went on to say:

> *By war's end, Brown had established Connor Brown Cadillac in Fort Lauderdale, built a branch of Nolan-Brown Motors on Bay Harbour Island in Miami Beach and relinquished the Trail Pontiac franchise in Miami.*
>
> *In 1965, Connor Brown was the largest Cadillac distributor in the Florida Region....That year Cadillac terminated all of its distributor agreements leaving Connor Brown as the Cadillac dealer in Jacksonville, Miami, Miami Beach and Fort Lauderdale.*

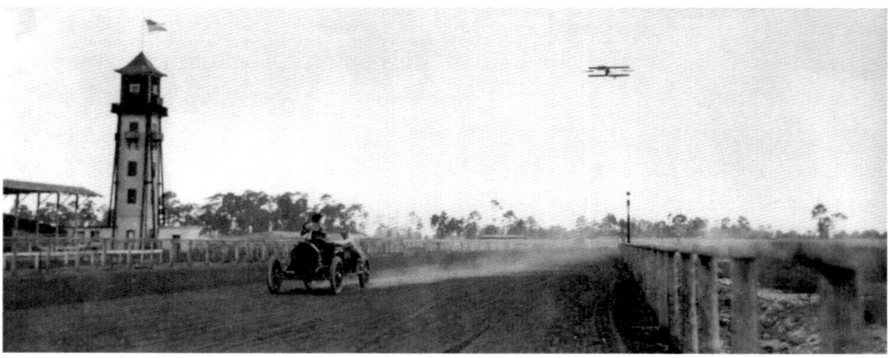

Claude Nolan's Cadillac beating an aeroplane at Moncrief Fairground. *Courtesy of Jacksonville Historical Society Collection.*

On November 22, 1976, Connor Brown died, and Claudia took over operation of Connor Brown Cadillac until 1979, when she sold the company. Jack Helmick, the son-in-law of Connor Brown and who had been a part of the dealership since 1965, became president of Claude Nolan Cadillac Inc. in 1977, after Brown's death.

I spoke with Marc Helmick, Jack Helmick's son and Claude Nolan's great-great-nephew. He is now general manager of the present-day Claude Nolan Cadillac on Southside Boulevard. On a beautiful, crisp November day, we talked about cars and the history of his very interesting family and business.

Helmick told me the story about Claude Nolan's drive down the railroad tracks to Key West and his race against an aeroplane. But there was also something else interesting that Claude Nolan used to do. He would write letters to customers and friends and, via American Express, have the letters forwarded until they had circumnavigated the globe, collecting postmarks. Helmick said, "Some of our older customers find these letters all the time in their attics or when they are cleaning out closets, and they bring them in for us to see."

He showed me one of the letters that he had framed. It told of the Hindenburg tragedy. It was amazing, like a time capsule. Nolan sent it from Jacksonville, and it made its way all through Europe and Asia before it came back to a recipient in Jacksonville. It apparently was quite a thrill for those who received such a special gift.

Although we couldn't put our hands on any specific photographs, Marc shared his story about the Claude Nolan dealership as we sat in his office and '60s rock music played softly on the intercom. As I was leaving the business, I saw in the showroom a beautiful antique Cadillac. It was a 1908 Cadillac, which Claude Nolan had sold to Mr. O.P. Woodcock, a prominent builder in the area. (He was the contractor of a house built around 1926 in San Jose on San Carlos Road that my sister had owned in the early 2000s.) Claude Nolan re-acquired the car from Mr. Woodcock many years later, and except for it having been repainted to a red color, it was in original condition and now sits proudly in Claude Nolan Cadillac's showroom.

As I inspected the beautiful machine, I realized that this automobile had been constructed a mere twelve years after Einig's "motor wagon" was mentioned in the newspaper for the first time and only three years after one of Nolan's Cadillacs had raced the aero plane. For a brief moment, I felt a momentary connection with the distant past—the past in which horseless carriages were amazing futuristic marvels, flying at

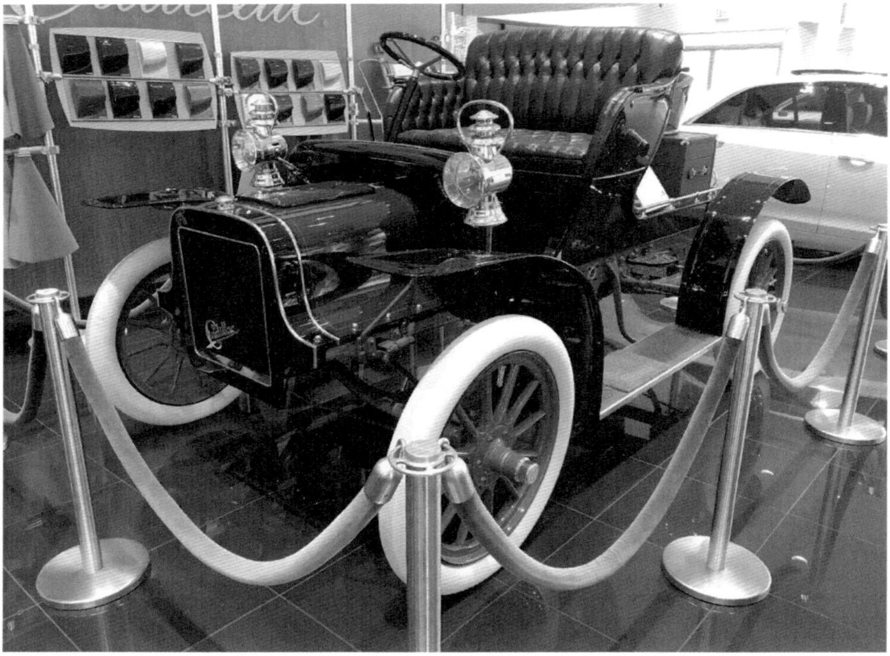

O.P. Woodcock's Cadillac in the Claude Nolan showroom. *Author's collection.*

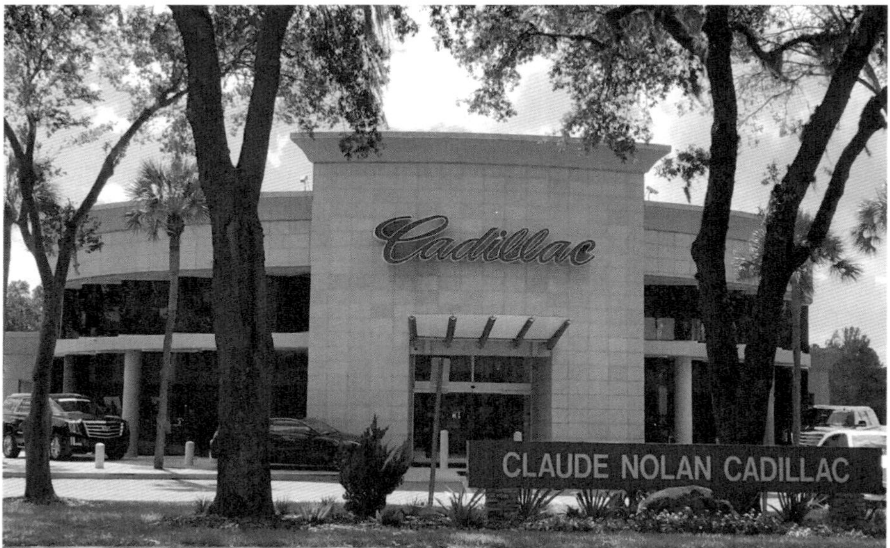

Present-day Claude Nolan Building. *Author's collection.*

breakneck speed down roads and boulevards of a developing metropolis. In that moment, I could understand the thrill that must have fluttered many a heart back when Jacksonville's culture was just on the cusp of something wonderful: a world where a conveyance didn't require a horse—the world of the automobile.

Duval Ford and Beyond: The Scott/McRae Automotive Group

There is no doubt Mama wanted me to drive more than I did. I was just plain scared even though I took driver's education at DuPont High School. Finally, my boyfriend and I headed for Oaklawn Cemetery where he patiently taught me to drive over several Saturdays, and I tried hard to avoid embarrassment because I was supposed to be a smart girl.

I was learning to drive "Rattles," Mama's old, white Ford station wagon in which she carted loads of kids to the beach every weekend. All went well at Oaklawn until I ran into a palm tree with my erratic gear shifting. Thank goodness Rattles looked pinged enough that it made no difference.

—Lynne Raiser

On a brisk, lovely spring morning, I made my way to the Five Points Office Building in Riverside, a beautifully renovated building that houses the SunRay Cinema and the offices of Mike Shad. I was hoping to learn what I could about how the Ford Motor Company was involved in the automotive history of Jacksonville.

I took a seat in the bright space across a large desk from Shad, and we began to discuss his life in the car business. Since our time was limited, he directed me to an article written about him some years ago by Charlie Patton and published in the *Times-Union* on January 8, 2017. In it, Patton chronicled Shad's time in Vietnam (September 1966–November 1969), as well as his considerable health issues as a result of his service there. He also included much about Shad's career in the automobile business:

> *He then went to work for his father-in-law at what then was Southside Ford, going through a two-year training program that included working as a used-car salesman.*

> *In 1985 he bought the business and two years later moved the dealership, renamed Mike Shad Ford, to Blanding Boulevard. While he was earning a master's degree in business administration in an executive program at Jacksonville University, his classmates made it a project to create a series of advertisements with Shad as company spokesman.*
>
> *He sold the dealership in 1997 to Auto Nation, and stopped being a television spokesman, though he stayed on as an executive with Auto Nation for several more years…*
>
> *In 2004, he bought and renovated the Five Points office building that also houses Sun-Ray Cinema.*
>
> *In 2005, he agreed to buy Bozard Ford Lincoln in St. Augustine from his wife's uncle and cousin, neither of whom was healthy enough to continue to operate the dealership. Shad bought it with the promise that he would sell the business back to the grandson of his wife's uncle once the grandson was old enough and experienced enough to run it. Shad said that sale will probably happen next year* [2018].

When it came time for me to leave, Shad showed me a photograph of his father and future father-in-law working under the same roof at Duval Motor Company, back in the late '40s. Shad also had a display of old-time automotive tools set out on a table near the photo. He looked at them with obvious pride, and I felt honored that he had shared these family treasures.

I needed to do some backtracking to understand the connection between all of these "Ford People." I got my answers some months later. It was another glorious day when I met with Diane McRae Graham at The Brick in Avondale, and the hour flew by as we ate our salmon and kale salads and talked about the car business. Diane was a very logical person to interview given how she and her family have been part of a powerful automotive force in the Jacksonville area: the Scott/McRae Automotive Group–Duval Ford. Her grandfather, her father, her husband, she and, finally, her sons have been part of the Duval Ford family and beyond for more than one hundred years. It seems that once her grandfather got into the business, all the members of the family became part of it as well.

According to a pamphlet published to celebrate the eightieth anniversary of Duval Ford, "Celebrating 80 Years of Tradition and Excellence," it was in 1916 that the new Duval Motor Company began selling Model Ts with a handful of employees. In 1922, a man by the name of Walter Alexander McRae, Diane Graham's grandfather, joined the Ford Motor Company. She told me that he had attended University of Florida and played baseball

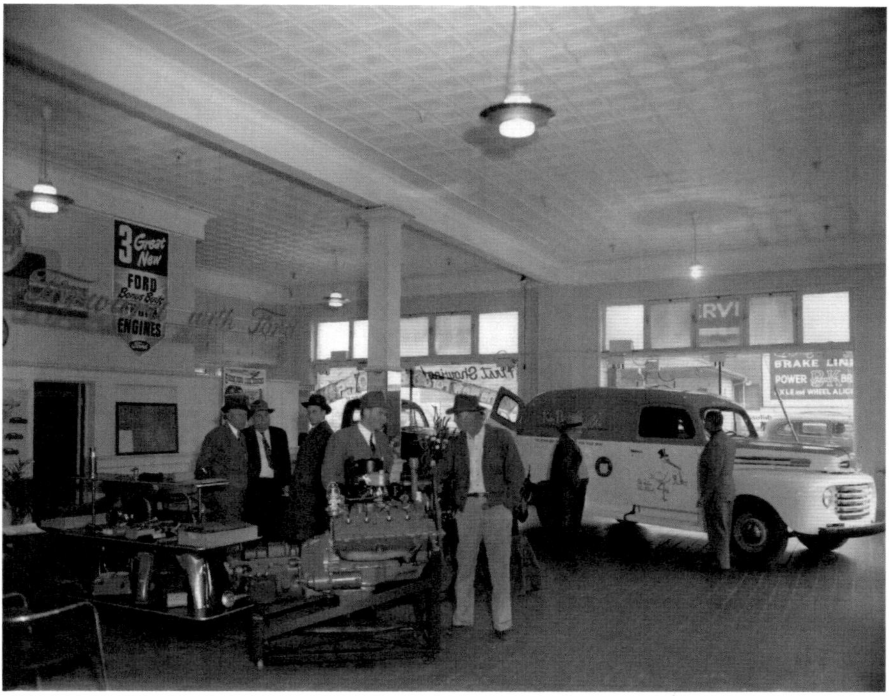

Duval Motors showroom. *Courtesy of Florida State Photographic Archives.*

while he was there. When he married Jennie Lynch Cross, they settled west of Jacksonville, where he worked in a mechanics shop. When he was not getting ahead enough, the family sought its fortune in Jacksonville. McRae went to Duval Ford to get a job.

"They had him unload boxes from the train cars," Graham told me. "When someone came back to check on him, they asked him why wasn't he working. He said he had already finished the job. They hired him permanently right away."

Graham often called her grandfather a "P.T. Barnum" type, but not in a hustler sort of way, but rather because he was "bigger than life." She said that whatever it was, her grandfather (whom they called "Fodder") threw himself into it. She often imagined him going to the bank as a young man and having the audacity to ask for money to buy Duval Motors—and having the charm and charisma to actually get a loan!

Graham told me that her grandfather had raced cars on the beach. He was also an avid sports fan. If it had a winner, he loved it—football,

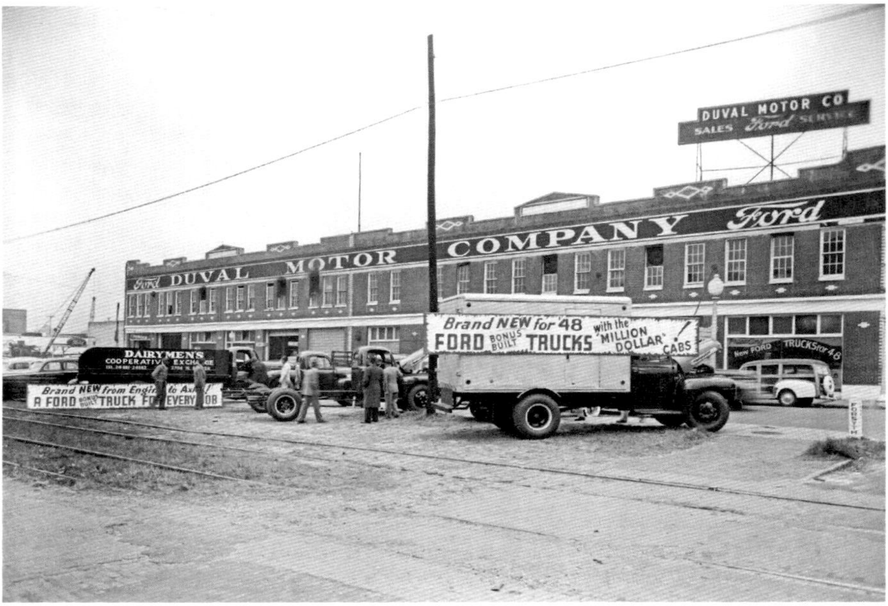

Duval Motors. *Courtesy of Florida State Photographic Archives.*

baseball, harness racing and even boxing. He and his wife would host boxers traveling through town and let them stay in their home. It is no surprise that McRae Sr. was one of the principals in the construction of the Gator Bowl in the mid-1940s, and he was also one of the original founders of the Florida Gator Boosters. The pamphlet said that at one time in Jacksonville, the only place to buy tickets to the annual Florida-Georgia football game was at the dealership.

One of the most important elements of McRae Sr.'s legacy is that he was all about loyalty and taking care of his people. The anniversary pamphlet relates how in 1926, at the Christmas party, Mrs. McRae served her homemade eggnog and fruit cake. Ernie Kopp Sr., who would eventually become a VP and partner, said a prayer, and then all twenty-five employees received a five-dollar bonus. It was during this period that the "tradition of family" became part of the Duval Ford paradigm. Mrs. McRae's eggnog recipe is still used for the Christmas parties these days.

During the Depression, when things were tight, McRae Sr. never missed a payroll, even taking out loans to be sure that everyone was paid on Friday. When civilian cars were not being made during the World War II, they sold charcoal, dental chairs, pots, pans and fertilizer. If they could buy it, they would resell it to keep the business going.

It also said in the pamphlet that McRae Sr. hired parolees coming right from Raiford State Prison. He said, "I'll take your murderers, but don't give me any thieves. Murderers change. Thieves don't." It was said that many of these men not only earned his trust but also stayed for the rest of their working lives.

In 1938, Walter McRae Jr., Graham's father, joined Duval Motor Company as a part-time employee. He eventually served in World War II for two years and then graduated from the University of Florida, joining the company full time in 1946. In the next year, McRae Jr. was able to persuade Jack Scott, a fellow UF graduate, to join the business. Together, they would form a partnership at Duval Motor Company, and then in 1980, the Scott-McRae Group was formed for the businesses they had created and built during their time as partners.

McRae Jr. continued the same "tradition of family" by being instrumental in spearheading a plan by which employees could get more than just a watch when they retired. He created a profit-sharing plan so that retirees could have money to live on when they left the company.

Jack Scott was also a force of nature, although he is generally characterized as being a "quiet leader." Scott had distinguished himself in the service during World War II as a navy pilot. One great story about Scott was related in a DVD lent to me by Steve Rawlins, CPA. It was said that Scott never had to buy a drink at an officers' club while he was in the service because there was always someone there who had been rescued by him.

Diane Graham also told me that it was Jack Scott who eventually became the father-in-law of Mike Shad. Both had worked together at Southside Ford before Shad bought it. Shad's wife's cousin is a Bozard, so it was only natural that the Ford line of products would move south to St. Augustine.

In an article written for the *St. Augustine Record* on November 10, 2005, Fred Bozard III related:

> *"It was a natural for Mike and me to get together," Bozard said.*
>
> *"We've been together for a long time. And the family's been in the car business since about World War II. This is what we do."*
>
> *Bozard Ford was founded in 1949 by Fred Bozard Jr. and Josephine Bozard. Bozard Ford is possibly the second-oldest continuously run family business in town. The Alligator Farm on Anastasia Boulevard has been continuously owned since 1937.*

As our luncheon continued, Graham began to talk about her husband, Henry Harris Graham, who was also called "Tip." They had been married for thirty-five years when he died from ALS in 2010, the same year that her father died from the effects of Alzheimer's. It was a very rough year. She lit up when she talked about Tip, though. She told me that he had majored in communications at the University of Florida and then earned a law degree. They met through mutual friends and had an incredible life together:

> *Tip was a lawyer when we got married. He went from one day trying a case to the next day moving over to Duval Ford bringing his legal pad to take notes on all aspects of the car business. He brought a very different energy to the equation. His parents were a doctor and a banker, and he really was an introvert who had to become an extrovert. He also came up with* Bumper to Bumper, *the company's newsletter, thanks to his communications background. But his great gift was mentoring. He was always planning events and company parties, and even when he was sick, he never lost his sense of adventure.*

Two of their three children are in the car business. Hampton Harris Graham and Alexander McRae Graham now run all of the businesses of the Scott/McRae Automotive Group–Duval Ford, Duval Honda, Duval Acura and Subaru of Gainesville. Their daughter, Katie, has chosen another path.

Graham said that each generation contributed to the next. Walter Sr. was the trailblazer of the family. Walter Jr. brought in the use of psychology to figure out what motivates people. Graham brought the more advanced communications and mentoring skills. The "boys" (as she referred to them once and then corrected herself to say "men"), Hampton Harris Graham and Alexander McRae Graham, brought with them the idea of streamlining the car-buying process. They are trying to get the experience down from an all-day event to two hours. Each generation gives a new twist on the car ownership process.

In the final analysis, Graham said, "The one thing that I think is so appealing [about the automotive business], despite the risks and the pressure, is that you can determine your future. You can manage your time. You don't have to move if you don't want to, but your company is moving. The greatest thing is the ability to put down roots. You control your destiny and you can stay connected."

We parted friends and may just have to do lunch again.

North Florida Lincoln and Key Buick

I was in X-ray training at the old Duval Medical Center making $75 a month. Still living at home and not paying for rent or food, I went to Key Buick on Philips Highway and met with Jerry James. I explained my situation and he financed my car there and my monthly payment was $50. That left me enough money for gas and insurance. I was thrilled! My first car and I did it all by myself! After I graduated I had black racing stripes painted on the sides and a black roof to make it look sportier.

—Darian Hulslander

Every few years my parents would get a new Buick from "Key Buick" when they were on Philips Highway. This had to be back in the 1960s and 1970s. When you would purchase the Buick, the salesman would send you a dozen roses.

—Margaret Rose

As often happens when one writes a book, the writer acquires information in a roundabout way. So it was for me when I was researching the Lynch family. Late in the month of April, I had the opportunity to speak with Bill Lynch, the third generation of the Lynch family. He is presently the head of North Florida Lincoln Mercury. He had invited me to his office located on Southside Boulevard, and on that day, North Florida Lincoln Mercury had just finalized the purchase of the Key Buick franchise across the street. It was a happy day all around.

Bill Lynch, who graduated from Bolles in 1977, began his career in the car business in high school—first by washing cars and then by learning automotive mechanics at the family business. This was a family tradition, it would seem, since all younger Lynches learned all aspects of the business from very early ages in preparation for taking the reins of the company someday.

In earlier times, it was a common practice for dealerships to provide twenty-four-hour service, especially for vehicles that were used by businesses during the day. The showrooms might be closed, but the garages were not. "It was really hard work in the garages," Lynch said, "especially in the South. We knew of a Birmingham dealership that actually suspended work in the garages at about 3:00 p.m. because it was just too hot in there for human beings. When things cooled off, the garages would open again for the evening. Of course, when air conditioning came along, it was no longer necessary to shut down like that."

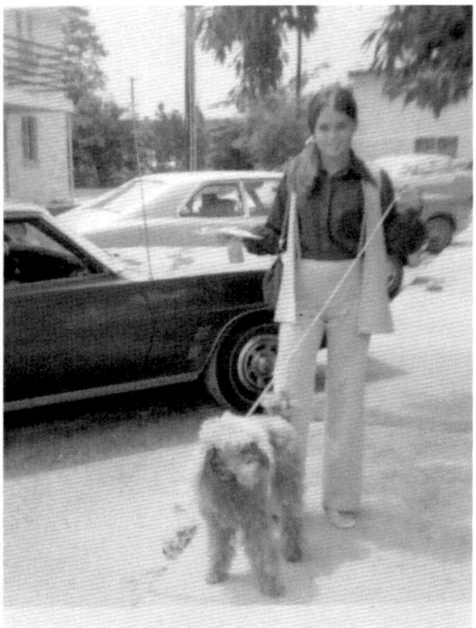

Darian Hulslander with the car she bought all by herself. *Courtesy of Darian Hulslander.*

He told me that the success of the business over so many years was based not only on good service and quality products but also on treating one's employees well. "This family business keeps its people, and they seem to love it here; loyalty is key, and we have many employees who have been with us thirty or forty years." He also said that dealerships really entrench themselves in a community—they sponsor teams, contribute to charities and involve themselves in community activities.

As we ended our conversation, Lynch encouraged me to talk with his father, who was still very much a part of the business. So, a few weeks later on a blustery May day, I found myself at the Lynch Management Company in Riverside. The ninety-eight-year-old Ransom Buffalow–designed home in which we would talk stood magnificently across the street from the choppy St. Johns River. It was a lovely day and a beautiful house.

I walked up the exterior stairs to the front door and rang the doorbell. Soon, I was escorted to the upstairs office of Mr. Lynch. He sat at a large desk next to a window with a commanding view of the river and the city. Mr. Lynch rose, greeted me warmly and then graciously showed me to a table where we could more comfortably talk about his sixty-four-year career in the car business.

It was a bittersweet day. The beauty of the building and the magnificence of the weather filling the breathtaking view contrasted sharply with the knowledge that Mr. Lynch's wife of sixty-four years was, at the moment, under hospice care. It had been a difficult time of late for the Lynch family. Still, he made time to conduct an interview, and that spoke volumes to me about the kind of man Mr. Lynch is.

"Some things get to be a blur," he told me early in the interview, but I found him to be exceedingly clear and helpful. His success in business and life became abundantly obvious to me as our time together ticked by.

The story of the Lynch dynasty began with Mr. Lynch's father, Hal Lynch Sr. He was born on a farm in Georgia, and in the 1920s, after the boll weevil

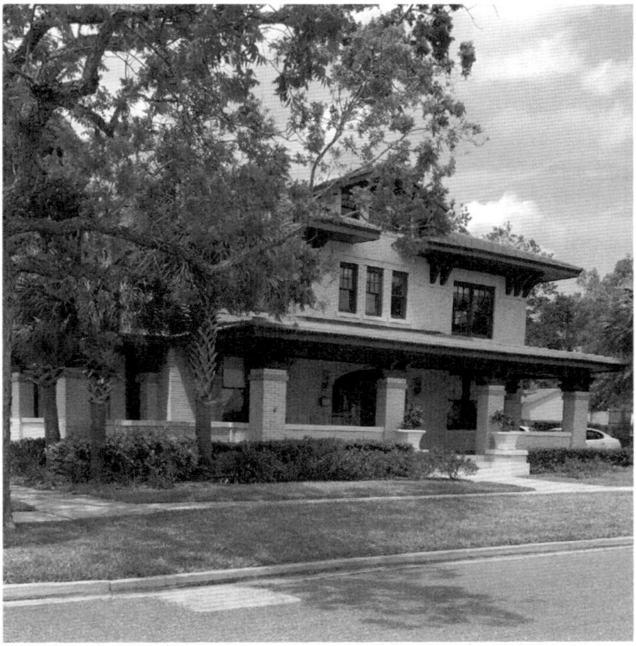

Right: Lynch Headquarters. *Author's collection.*

Below: Lynch Motor Ball Team. *Courtesy of Bill Lynch.*

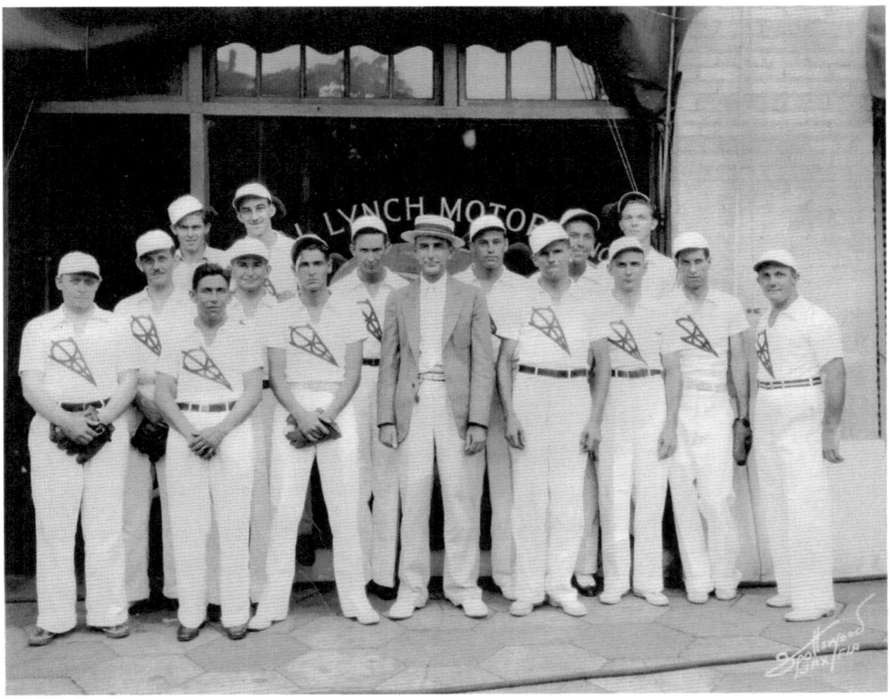

had decimated the family's cotton crop, Hal Lynch Sr. sought his fortune in the big city of Jacksonville, Florida. He worked at an assembly plant on Wambolt Street near where the Mathews Bridge is now, and with simple deduction, I realized that this was *the* Ford assembly plant (discussed later in this book).

Lynch Sr. worked in the plant for a while, and then he worked as a truck salesman, selling Reo Trucks. It wasn't long before he decided to go into business with Bill Wright. Together, they formed Lynch/Wright Motors, which began in 1934 and was located at Duval Street. In 1935, F.F. "Tye" Davidson bought Wright's interest in the company, and Lynch-Davidson Motors came into being.

In an article published on jacksonville.com, it noted that Davidson also came to Jacksonville from Georgia. In the early 1920s, Davidson left Shadydale, Georgia, and he, like Lynch Sr., went to work at the Ford factory on Wambolt Street. Davidson, who already had a degree from the University of Georgia, rose quickly up the ranks. Within three months, he had risen from the assembly line to become manager of the plant.

Life was not easy in the car business during World War II. Every dealership had trouble getting new cars because of the war effort. During these lean times, Lynch Davidson got into rental cars, and the company began to help with the building of Camp Blanding by supplying automobiles and trucks for transporting personnel and supplies. However, by 1943, Hal Lynch Sr. was able to create North Florida Lincoln, according to company brochures.

Lynch Sr.'s son, Hal Lynch Jr., joined the business in 1953. That year was quite momentous for Lynch Jr.—he left the service, got married and went to work in the family's car business. Lynch Jr. met his future wife, Frances Thompson, in Honolulu in 1951 when he was stationed in Hawaii while serving in the U.S. Coast Guard. Frances had been born in Honolulu and had even witnessed the attack on Pearl Harbor. She was quite the beauty, an accomplished musician and a hula dancer. The couple was married in 1953, and they settled in Jacksonville. In 1964, Lynch Jr. took over for his father.

At the time when Lynch Jr. started his career in the car business, passenger cars sold for $1,895. Most people wanted V8 engines, and most shunned the straight 6 engines. He also mentioned that his job was to sell trucks, but back in the early '50s, it was hard to sell trucks. "People just didn't seem to want them the way they do these days," he said.

According to a special article titled "Blue Ribbon Celebration," written for the *DRIVE* section of the *Florida Times-Union* on January 17, 2014, Lynch Davidson Motors flourished. Then, in 1968, it moved to Atlantic Boulevard

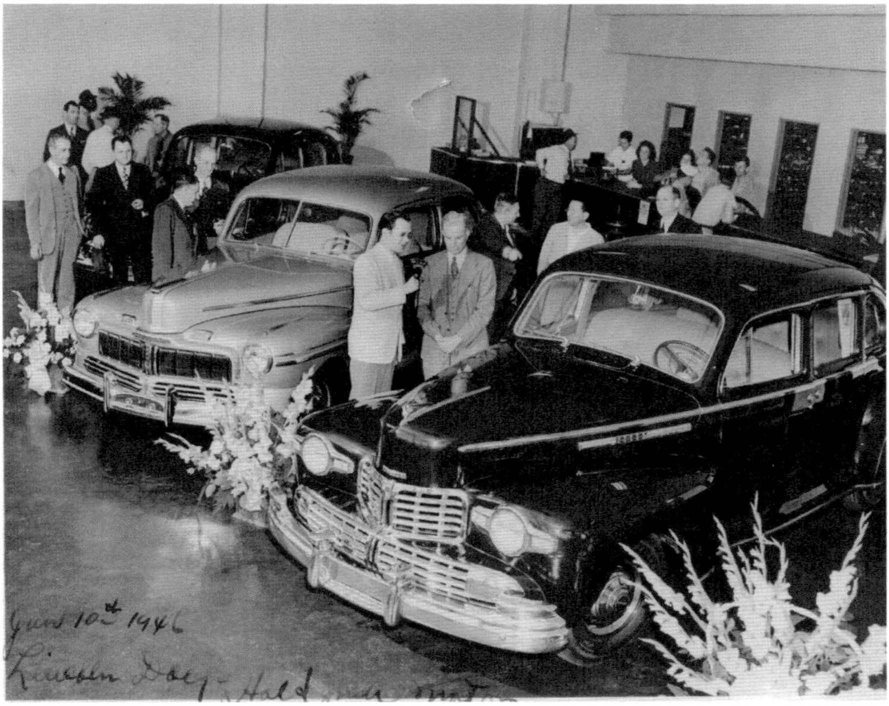

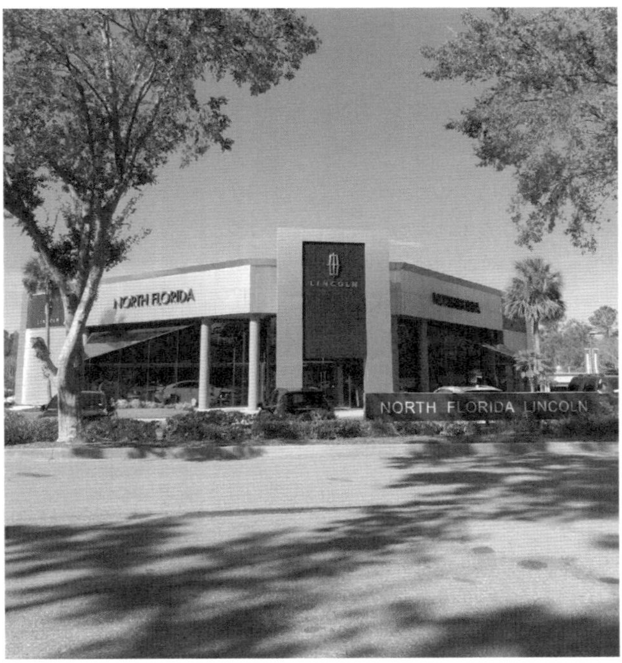

Above: Lincoln Day. *Courtesy of Bill Lynch.*

Right: North Florida Lincoln today. *Author's collection.*

across from Regency Square Mall. In 1969, Davidson bought out Lynch's interest in the company, and it became Davidson Ford. The Lynch family went on to buy North Florida Mercury Motor Company on Forsyth, one on Riverside and another on Philips Highway.

In 1981, F.F. "Tye" Davidson died, leaving his son, Mike Davidson Sr., to take over the business. The business was then named Mike Davidson Ford. Mike Davidson Jr. took over in 2004 after his father died, and in 2015, the dealership was purchased by Ashbury Automotive Group, which owns Coggin. It is now called Coggin Ford.

When Hal Lynch Jr. and I ended our interview, we shook hands, and as I prepared to leave down the upstairs hall, I noticed amazing pictures lining all the walls. I came to one that made me stop and take notice. There was a very large photograph of Hal Lynch Sr. as a young man, and he appears to be deep in a conversation with Henry Ford himself. Such an amazing photograph!

Sadly, the very next day, Frances Thompson Lynch passed away, certainly leaving a great hole in the heart of the family so much a part of Jacksonville's automotive history.

KEY BUICK

According to the Key Buick website, the Key Auto Company was started by Thomas W. Perry in 1960, and his Buick dealership was located on Philips Highway. When Perry Sr. died in 1979, his son, Thomas W. Perry Jr., became president of the company. In 1985, the business moved to its current location on Southside Boulevard. Maureen Perry Burnett, Perry's sister, was a certified public accountant in Knoxville, Tennessee. In 1995, Perry asked her to come to Jacksonville and serve as business manager for Key Buick. In May 2008, Burnett took over as president of the company.

According to Roger Bull of the *Florida Times-Union*, Lynch Automotive Group, owners of North Florida Lincoln, reached an agreement on July 2, 2015, to purchase of the Key Auto Company, owner of the Buick, GMC, Hyundai and Genesis dealership on Southside Boulevard. And now the neighbors are family, so to speak. The past gives way to the present.

THE FORD ASSEMBLY PLANT

No history of automobiles in Jacksonville would be complete without mentioning the old Ford Factory located on Wambolt Street. On an ordinary summer morning, my husband and I set out on a grand adventure to find this legendary edifice. Armed with copies of two wonderful books—*Reclaiming Jacksonville*, written by Ennis Davis and Robert Mann, and *Jacksonville's Architectural Heritage*, written by Wayne Wood—my husband and I set out to find the long-lost Ford assembly plant, which had been part of the Jacksonville economy in the mid-1920s and into the 1930s.

We crossed the St. Johns River by traveling over the Isaiah D. Hart Bridge, and we exited at the Sports Complex to get to Talleyrand Avenue. We followed the commands of our GPS and eventually happened upon Wambolt Street, where an opened, chain link fence with a "NO TRESPASSING" sign stopped us. What appeared to be a huge, derelict building loomed beyond about a football field's length away. Clearly, the bones of the building matching the old photographs in our books rose ominously before us.

Carefully and bravely, we drove past the unmanned gate of what looked like a construction/storage facility and slowly drove about the old complex, trying to imagine the tremendous activity that must have occurred in this automobile factory that once flourished on this property more than ninety years before.

According to the chapter on the Ford assembly plant in *Reclaiming Jacksonville*, it was back in the early part of the last century when Jacksonville was one of many cities chosen for one of the satellite plants of the bigger Ford factories located in Michigan:

> *Similar facilities were constructed in Charlotte, Chicago, Cleveland, Copenhagen, Dallas, Dagenham (England), Edgewater, Kansas City, Long Beach, Memphis, Mexico City, Philadelphia, Seattle, St. Paul and Toronto.*
>
> *The arrival of the world's largest automotive manufacturer was a sign of Jacksonville's growing economic importance.*

The chapter goes on to say that in 1923, Ford bought a ten-acre tract of land, formerly the Bentley Shipyards, from the City of Jacksonville for $50,000. Here, for $2 million, Ford would construct a 115,200-square-foot assembly plant designed by Albert Kahn.

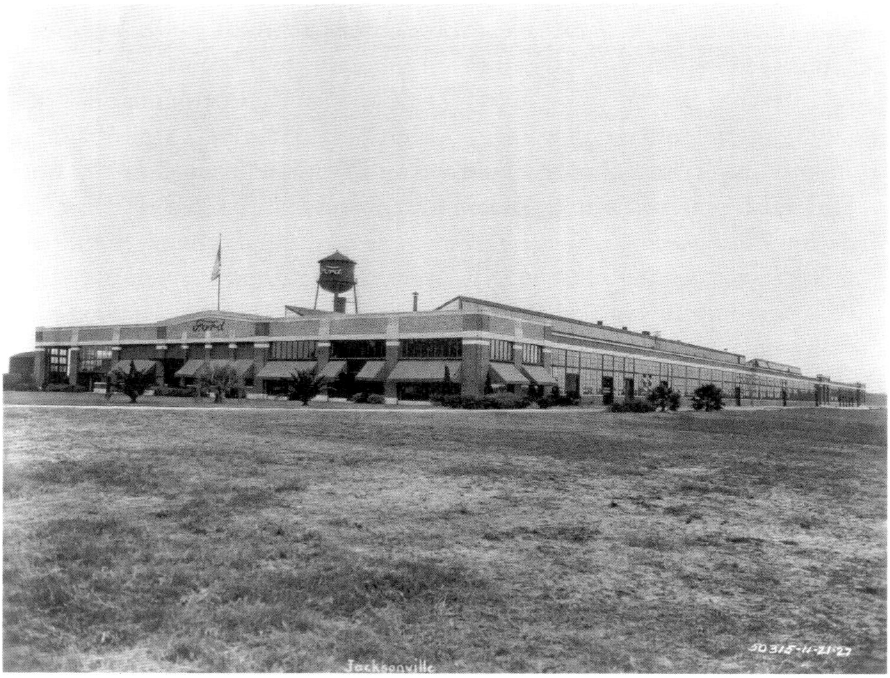

Ford assembly plant, 1927. *Courtesy of Ford Motor Company.*

In Wayne Wood's *Jacksonville's Architectural Heritage*, he noted that the Ford Motor Company Assembly Plant is "one of 1,000 buildings designed for Henry Ford by Albert Kahn, an internationally recognized industrial architect." Wood noted the plant was originally 200 feet by 576 feet, and within a year, it was lengthened to 800 feet. The building had "enormous skylight panels, which extend several hundred feet in length and provide natural lighting and heat to the interior. Also the sides of the building are made largely of glass. The west façade of the plant features a parapet of basket-weave and herringbone patterned brick, and the horizontal emphasis of the building is accentuated by a masonry band that runs completely around all four sides."

Earlier, I had found an article that appeared in the *Florida Times-Union* on August 29, 1924, announcing, "FORD OFFICERS MOVE IN PLANT":

> *The completed establishment including reinforced concrete docks will represent as expenditure of more than $2,300,000. From an industrial standpoint the plant is one of the greatest achievements in the history of Jacksonville. With the final completion of the docks and operation of the*

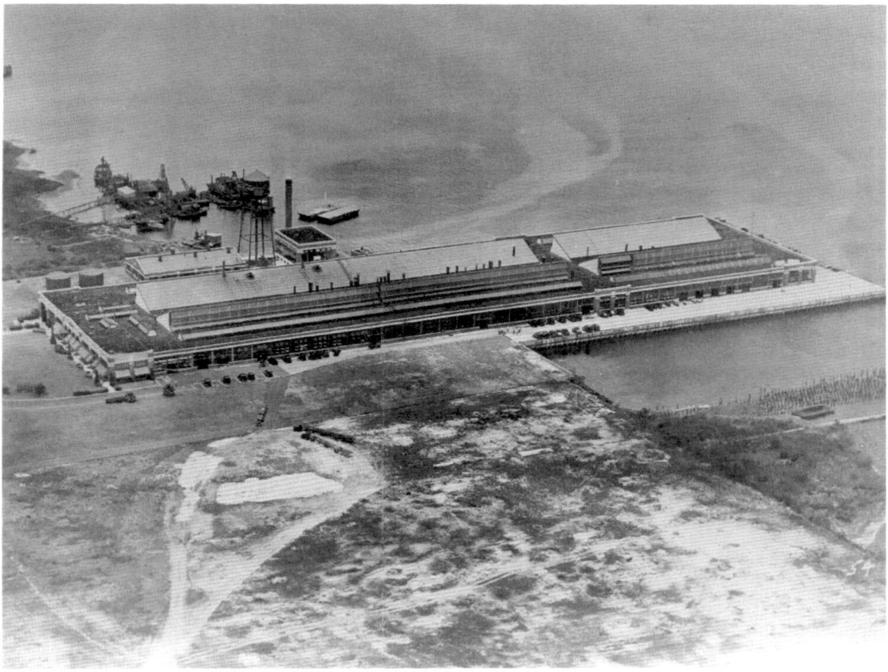

Aerial view of Ford assembly plant. *Courtesy of Ford Motor Company.*

> *entire plant system, Jacksonville will become the principal shipping center for the Detroit factories and hundreds of cars will be sent through the local plant, Mr. Rowman* [manager] *stated.*
>
> *According to tentative plans a general celebration and official housewarming will be observed when every department has been put into operation, with executives from different sections of the country and representatives from Detroit in attendance. Mr. Rowman stated that visitors would be welcomed at any time.*

Then, on November 26, 1924, production of the Model T began here in Jacksonville, Florida. According to *Reclaiming Jacksonville*, 125 automobiles were produced each day on average, but even that number was not enough to keep up with demand of Florida's "rapidly growing population."

On March 1, 1925, an article ran in the *Ford News*, a company publication, announcing, "New Jacksonville Plant Completed; Capacity 150 Cars Per Eight-Hour Day." The article noted that the new Ford assembly plant was "now occupied and in operation."

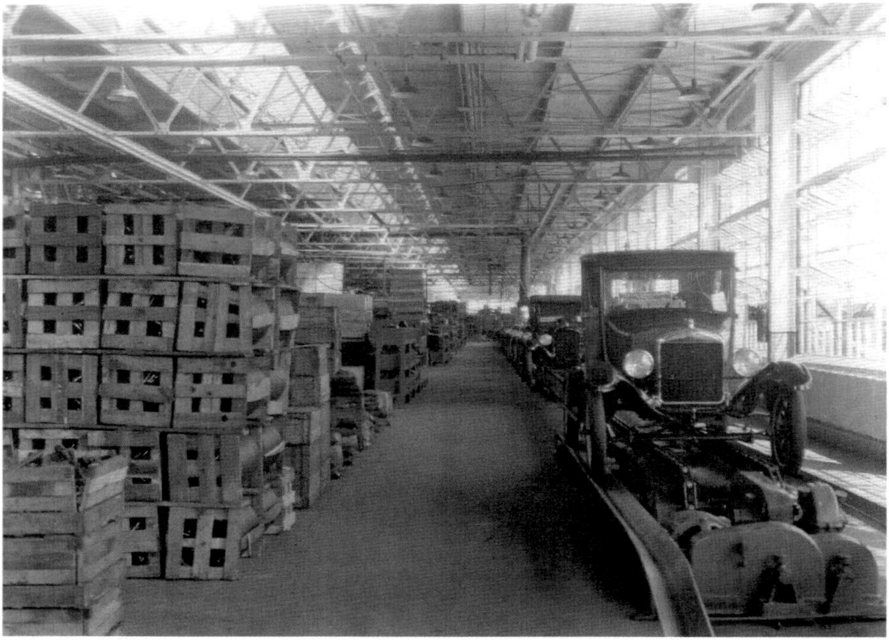

Assembly line in Jacksonville. *Courtesy of Ford Motor Company.*

Although some of the dimensions in the *Ford News* article differed a slight bit from the other sources, the numbers tended to be rounded appropriately and were not meant to be used for construction purposes. There were other tidbits of information, like the fact that 900,000 bricks were used in the construction of the building along with eighty thousand one-hundred-pound sacks of cement.

The Ford article also noted, "In addition to an improved body oven, the second of its kind in the United States, the plant contains a complete glass cutting and grinding unit." It went on to say that since the plant was built on the banks of the St. Johns River, it had excellent rail and water transportation: "A rail line enters the building on the side opposite the assembly line and another runs along the outside of the building extending to the dock of reinforced concrete jutting into the river."

The SS *Oneida* of the Ford fleet had already made two "calls" on the plant, bringing with it five hundred tons of cargo each voyage. The space vacated by the cargo was filled with "commercial cargoes," which I took to mean cars. These were then shipped to facilities in Gulf ports that "hitherto were unavailable to shippers using this port [Jacksonville]."

This demand meant that within a year, Ford would purchase an additional ten acres south of the plant for $114,000 so that the operations could be expanded. On November 13, 1926, the 50,000-square-foot expansion project was completed to make the entire complex 165,000 square feet. *Reclaiming Jacksonville* noted that by 1927, there had been 74,908 automobiles (mostly Model As) and 14,821 trucks assembled here in Jacksonville. At its height, the plant employed as many as eight hundred workers, producing 200 cars per day.

By November 1932, the Great Depression had been having a major impact on car production nationally, and Ford no longer needed the Jacksonville factory. When manufacturing operations ceased, the building was used as a parts distribution center for the company's southeast region until 1968. These operations were then moved to the Westside of Jacksonville and to Orlando.

The plant has been used by several companies for brief periods of time: World Cars (Toyota), a van conversion company, Boron Boats and a European car-importing company. By the end of the 1990s, the plant was being used for construction storage by Wood-Hopkins Contracting.

In the early part of the twenty-first century, a booklet was compiled called *The Ford Assembly Area Revitalization Study*. In it there was mention of the possibility of making a cruise ship terminal with loft apartments and businesses on the property. This project never came to be, and the once-bustling industrial space waits not far from the foot of the Mathews Bridge, where curious history buffs, like my husband and me, might wander about in dusty ruins. The silenced bones of the plant filled me with a melancholy that comes with my overactive imagination. We left as quietly and as undetected as we came.

AL SAGER MOTORS INC. VOLKSWAGEN

It was 1965 when I began my journey with personal automobiles. My mother passed away that year, and as a sixteen-year-old soon to turn seventeen, I inherited what came to be the symbol of teenage independence: a 1953 Plymouth Savoy, two tone with a turquoise body and a white top. It even had a small fan on the dash. The heater worked well, and the only air conditioning was a rolled-down window. That car opened the world of cruising, surfing and submarine races at the drive-in theaters. All teenage boys' delights!

> *In 1967, my dad purchased the only new car I ever had the opportunity to drive at the time…a brand-new, navy blue Volkswagen Beetle! From that point on, I ended up later with a '68 Beetle, and the first one I ever purchased on my own was a '73 Beetle.*
>
> *My fascination for the independence of having a car had waned tremendously by that time, though. I guess I was growing up, and the novelty just was not the same. From that point on, I purchased many cars just to satisfy my tastes for each decade there.*
>
> *—Jackson Lanehart*

As we sat in the lovely beachside home of Susan Sager Stein on a bright, cheerful day in May, we began a conversation about her family's role in the automobile industry in Jacksonville. We started by talking about her grandfather Alfred "Al" Sager, moving to her father, Abraham Elliot "Sonny" Sager, and then touching on the years with her then-husband, Bob Kogan. Her family most certainly has been an integral part of Jacksonville's automotive business.

It all started with a used car company called Al Sager Motors, located at 15 West State Street sometime in the '30s. She said that they moved their operations to a Monroe Street location, where they began selling new imported autos. "It was mostly English cars at first—MGs and Austin Healeys—but I'll never forget the wonderful Mercedes my father brought home one day. It was something else, with those gull wing doors that swooped up when they opened. And there was even a Citroen, a French car I remember."

Eventually, Sager Motors became the first Volkswagen dealership in the city, and even though there were many problems with being a VW dealership—getting inventory and complying with many and somewhat arbitrary German rules—it was during this period that Al Sager Motors really became successful. It was then located on Hogan Street.

The Sager family also became aware that there had been some grumbling at their synagogue that the Sagers were selling German cars once commissioned by Hitler, but her grandfather was a hardworking, pragmatic man who said that the war was over and a man has to have a way to make a living. That was that. The business finally settled on Kings Street on the Southside.

One of the greatest stories Susan told me was about how her father took her to New York. They stayed with relatives—who, by the way, made the best rice pudding ever—and then they drove all the way home in an MG that was

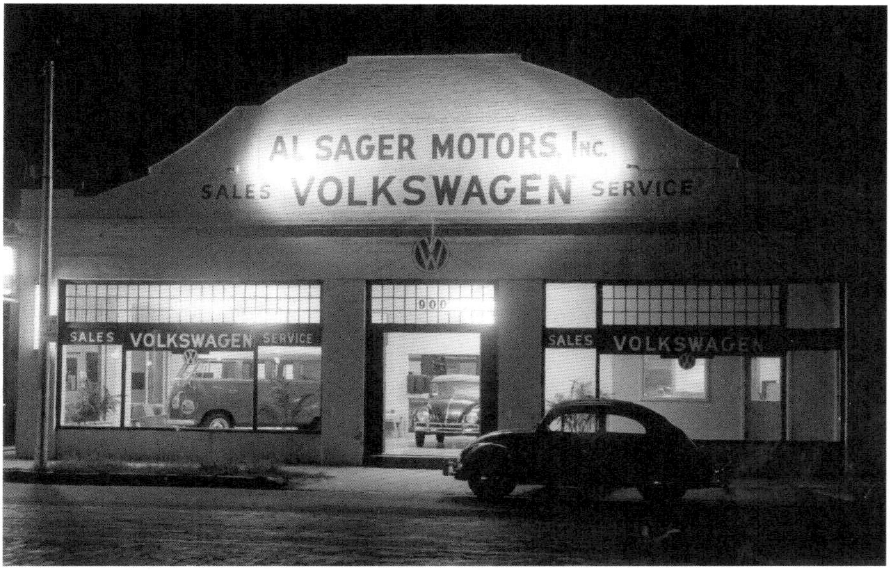

Al Sager Motors. *Courtesy of Bill Warner.*

soon to be featured in the showroom of the dealership. Susan was just in her teens when she and her father shared this special driving adventure.

She spoke of what a "great couple" her parents were—her father with his love of cars and her mother, Alice Biscow Sager, being the concertmaster for the fledgling Jacksonville Philharmonic, which would become the Jacksonville Symphony. Both were very proud of the other's accomplishments.

Susan also told me that she had an Alfa Romeo when she was a student at the University of Florida. It wasn't long before she was brought before the student council for having too many parking tickets. They actually took away her driving privileges for a while. "Everyone thought we were fabulously wealthy since we always had some really wonderful cars in the driveway, but the dealership owned the cars."

With her encouragement, I contacted her son, Mike Kogan, in order to flesh out some of the history with someone who was very fond of the business. It wasn't but a few days later that I met with Mike at the Lakewood Starbucks. It was so crowded and loud inside that we decided to sit out on the balmy patio and talk cars.

Mike admitted that he never knew Al Sager, his great-grandfather. "I am pretty sure he was gone before I was even born." He went on to say that

Alfred Sager actually got into transportation by trading horses initially. From there, he went into used cars and then into foreign cars, and of course, his son, Abraham Elliott "Sonny" Sager, joined him in the enterprise when he was able.

It was very evident that Mike Kogan had a very strong affection for his grandfather "Sonny" Sager. He often referred to him as "Papa" during our conversation. "My [maternal] grandfather and my uncle, Jeff Sager, owned 'the Place,' as we called it. And my father, Bob Kogan, was a general manager. When I was in junior high, I worked in the back, cleaning and prepping cars. It was there I got to know a Nigerian named Rudy who spoke the Queen's English as well as twelve other languages. We worked together, and he eventually he taught me how to drive a stick shift. That was, at least, until my father found out. I probably wasn't much more than twelve or so, and I apparently had been terrorizing the other workers."

Kogan also recalled how cool it was that Al Sager Motors sponsored Formula V race cars. "Papa and Jeff were nuts about racing. They used to go to Daytona and the Beach races. One of our drivers was Buddy Cox, and he won for us several times."

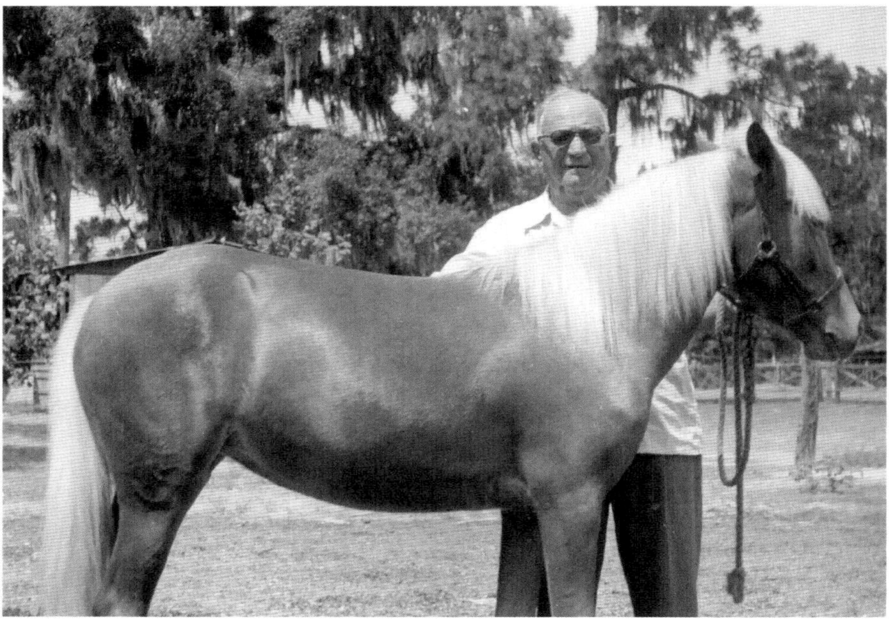

Alfred Sager with a horse to sell. *Courtesy of Bill Warner.*

In order to find out anything about these race cars, I went to talk with Bill Warner, the founder of the Amelia Island Concours d'Elegance, and here is as good a place as any to admit that Bill Warner was the impetus for my writing this book in the first place. Bill began his career as a car enthusiast working for "Sonny" Sager when he was sixteen. He drove the courtesy car and the parts truck, and Warner would "schlep" tires at the racetrack for Sager's race cars on the weekends.

I had actually met with Bill about six months earlier, and at that time, I discovered that Bill is just plain fast. On our first official meeting, before I even knew if I wanted to write a book about cars, he took me all over Jacksonville to see the sites of old dealerships. As he expertly drove his car around corners and quickly negotiated all the one-way streets downtown, he revealed his encyclopedic knowledge of everything automotive. He also made it absolutely impossible to take notes since it was hard for me to focus on my notebook, which kept sliding back and forth with each twist and turn of our journey. It is very easy to conclude that Bill Warner is pretty close to a perpetual motion machine.

By the end of the first interview, I was pretty sure that I could write this book. I knew it for certain when Bill let me sit in his Triumph TR-6, once owned and driven by Paul Newman. (Since that time, Warner has sold the car.) Getting into the driver's seat was a bit daunting since it proved a tight fit. Getting out the car proved almost impossible. Still, with much effort from me and Bill, I eventually emerged. It was quite a thrill to think of who all had been there before me.

When I went to Bill's office for a second interview, I was once again amazed at all the racing posters and posters for the Concours d'Elegance on the walls. I sat across the award-covered desk from Bill, who was busy at his computer. As we began this interview, he suddenly left the room and returned with a very old suitcase. In it were all sorts of papers, programs and photographs of Sonny Sager and his race cars. There was even a photograph of Jeff Sager and a VW that had been fitted with a Porsche engine. Bill said that he and Jeff Sager would challenge other guys driving down Philips Highway, and the other guys, thinking they could take a VW, were always surprised when the VW took off like a rocket.

There were two Sager race cars that were mentioned in the racing record books: the Mighty Mouse and the Mickey Mouse Specials. Both sported appropriate mouse decals, and crew shirts had the same embroidered "mouse" patches.

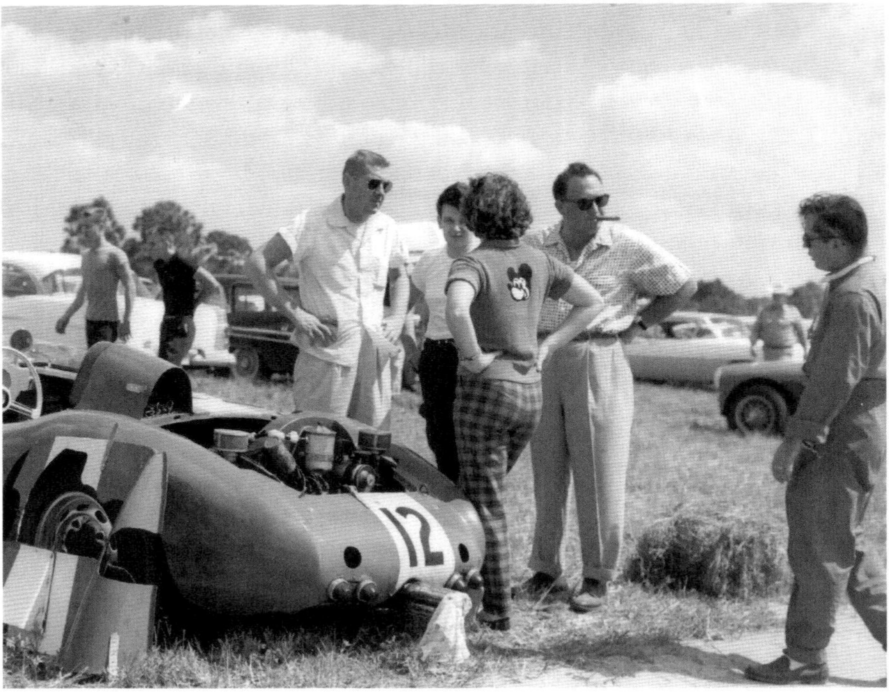

"Sonny" Sager with his race team: Jim Harcum, Karl Hloch, Marlene Harcum, Sonny Sager and Peter Laffe. *Courtesy of Bill Warner.*

After doing a little more research, I found a book titled *Vintage American Road Racing Cars, 1950–1970*, written by Harold W. Pace and Mark R. Brinker. In it, there is mention of both Sager cars. They were constructed in Jacksonville by Pete Laffe, Jim Harcum and Karl Hloch:

> *These two Porsche powered specials were built in Florida in the late 1960s. Constructed at Al Sager Volkswagen, the two cars wore distinctive mouse graphics. The Mickey Mouse was a simple car based on the VW floorplan with a Porsche Super Engine. The Mighty Mouse was more sophisticated with a tube frame, midengined Porsche pushrod engine and a highly modified Devin body.*

These two cars were also featured on the front cover of the April 1959 edition of a magazine called *This Week in Jacksonville*.

Our search through the suitcase produced more pictures. There were several of the cars being prepped for races, with the builders and the drivers

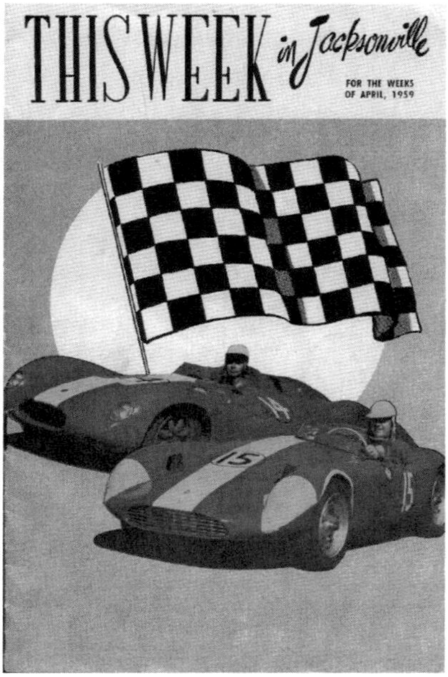

"This Week in Jacksonville," featuring Mighty Mouse and Mickey Mouse of the Sager Race Team. *Courtesy of Bill Warner.*

rushing about. And then there is Sonny Sager in the midst of it all with a cigar in his mouth—always a cigar in his mouth, I was told.

Kogan had told me that when "Papa" was of retirement age in the early '70s, he sold the dealership to his son-in-law, Kogan's father, in 1973. The business became Bob Kogan Volkswagen in 1974. Sadly, it wasn't too long after that the Oil Embargo came along—a time when most couldn't get enough gas to drive even a small, fuel-efficient car like a Volkswagen. Bob Kogan eventually had to sell the dealership, which he did, to Frank Griffin.

I have to admit that I have an affection for Al Sager Motors, one that I had forgotten until I started to research this book. The VW Beetle at my wedding—the first car belonging to me and my husband as a couple and the one that carried us away on our honeymoon—sports an Al Sager Motors license plate holder, a fact I had missed until I scanned my photos for my editor. That was a wonderful little car that got us off to a great start in life. For that I am quite grateful, and even now, seeing a Beetle drive by gives my heart a bit of a tug.

Frank Griffin Volkswagen

In writing this book, I have had many good and interesting times, but one of my favorites had to be the interview with Frank Griffin, aka Super Rabbit. Of course, when we got together at a Bob Evans Restaurant in Mandarin for coffee and tea, he was simply Frank Griffin. But he was a joy to talk to, and our time together went by too quickly.

He really got his start in the car business when he was serving in the U.S. Air Force in Germany around 1957. He was maybe twenty-two years old when he began selling cars to American GIs. He'd take their deposits and then had it arranged that they would pick up their new rides in New York after they got home and out of the service.

Griffin had every intention of remaining in Germany, but he followed—all the way to Jacksonville, Florida—the family of a young German woman who would become his first wife and the mother of his three children. They married and eventually settled in Jacksonville. He even had to have his German landlady send all his stuff over to Florida since he was only planning on a short vacation.

He worked for about a year at Lynch Davidson Ford, and then he went to Platt Pontiac to sell Edsels. He worked there for about fifteen years and then went to Fernandina as the general manager of a dealership there. When he returned to Jacksonville, he worked for Claude Nolan Cadillac for three or four years. He then went to work as a general manager in Tallahassee for Bob and Fred Drake. Finally, he felt that it was time to get his own franchise or dealership.

Bobby Kogan was the son-in-law of Sonny Sager (and the man to whom Sonny had sold his Volkswagen dealership). Kogan then sold the franchise to Frank Griffin in 1975. It then became Frank Griffin Volkswagen.

Griffin said, "I got to thinking about how successful all the guys were who did characters when they sold cars—Art Grendle in Orlando and Dick Balch in the devil suit in Seattle. So I went to this guy in San Marco who had a costume store, and before you know it, we have me in a rabbit suit with a big SR emblem on my chest. Children loved me. Women liked me, and men just moved out of my way. But it worked. A local radio station ran a survey that measured how recognizable certain people were around here, and I got 70 percent of the vote.

"And it didn't stop there. I did all kinds of events—even a few with the Southern Brick and Stone guy—Chicken Man, Dolph Ageloff [whose tagline was "Leave me give it to you again!"]. I even Rollerbladed in costume for a 10K run for spina bifida. Skating like that is not always easy. Bridges could be a problem. I had to be especially careful going over the cobblestone one in San Marco."

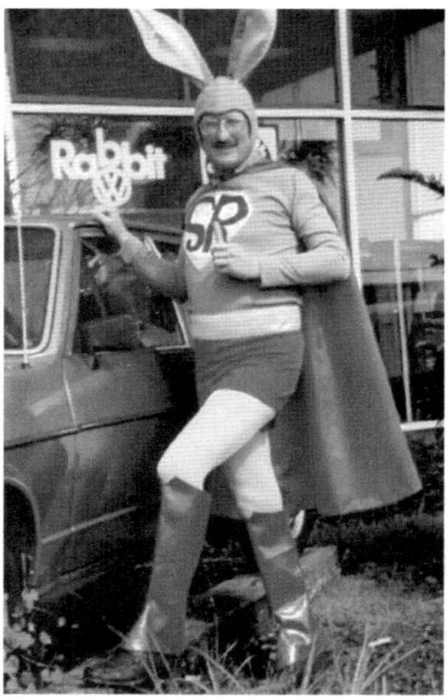

Frank Griffin, aka Super Rabbit. *Courtesy of Frank Griffin.*

Griffin has since left the costume and the salesmanship behind, although he does lease several properties that he owns. He seems to be enjoying life. "I have had a blessed life," he told me, and then we talked about things other than cars. We talked about our families and his great-granddaughters, whose pictures he showed me, and we talked about how things have changed.

He told me that in the beginning of his career, one had to be very good at math—all calculations had to be worked out on legal pads, and all accounting was done by hand. Then along came this adding machine thing, and Griffin really thought he was "uptown" when they got one of those. He even told me about a calculator that could do more than add. "I couldn't believe it! For $129, I could do all my calculations without paper or pencil! We also had this Selectric Typewriter that was so much better than the manual ones, and then Wang came out with this programmable model, so that our contracts didn't have to be retyped, over and over for each customer. Nowadays, we can hardly imagine what it used to be like."

And in the final analysis, there was one thing that Griffin wanted the world to know: by and large, car dealers really are honorable people: "Car dealers involve themselves in the communities that they serve, and they really do work hard to be fair to everyone. The secret to selling cars is to get the customers to come back when the time comes. You can't have that happen if you don't act fairly."

By the end of the interview, I understood why Frank Griffin had been so successful. His personable manner was warm and engaging. Believe me, I would have bought a car from such a guy!

Nimnicht Chevrolet

My dad bought me a new green stripped down, very basic, 4-speed on the column Chevy Nova for $2000. Boy, was I set! Freedom, freedom, freedom! They did this for me just before my sophomore year at Georgia Tech began [1968]. *They were very uncomfortable knowing that I was constantly hitchhiking, including commuting from Jacksonville to Atlanta. (They used to drop me off I-10 at I-75, and I would wait until someone picked me up.)*

I absolutely loved my Nova and drove all over Atlanta and Jacksonville. My friends nicknamed it "The Mouse" because of the small engine. I let my fraternity brothers borrow it. When I started student co-oping with

> *NASA in Houston on three-month stents, I could drive to Houston instead of the alternatives and have my own wheels when I arrived. Fun, fun, fun!*
>
> *After 184,000 miles, I finally traded it in for a new Toyota SR-5 hatchback (1976), which I promptly took off in with my girlfriend Mary on a 17,000-mile three-month road trip through forty states (camping, backpacking, adventure) and wound up settling down in Boulder.*
>
> —*Tom Spear*

In the year 1941, the *Jacksonville Telephone Directory* had a page that was filled with "Jacksonville Facts and Information." It said that 1,162,000 people lived within a 150-mile radius of Jacksonville, and that was a respectable number since the region the phonebook served consisted of parts of Florida, Georgia, Alabama and South Carolina, some 9 million people. In that same year, many industries were listed as thriving in Jacksonville. From lumber and millworks to warehouses and refrigeration plants, the Jacksonville area was a very busy place. The phonebook also said that "Jacksonville is one of the financial strongholds of Florida."

In the business section of this telephone directory, there was a listing for a new car dealership that was located at 123 Riverside Avenue. The dealership was called Riverside Chevrolet, and its phone number in 1941 was only five numerals long, 5-6980. (By 1955, the number would change to EL 4-0123 so that it included the 123 of its address.)

On January 15, 1941, Bill N. Nimnicht Sr. opened Riverside Chevrolet in the space where the Haskell Company building now stands on Riverside Avenue. He became the youngest Chevrolet dealer in the United States at the age of twenty-two. The present-day Florida Times-Union Building, which would be on the adjoining property, did not yet exist. In 1941, that property belonged to a bus station. A working firehouse was just down the block on Riverside.

On a beautiful March day some seventy-six years later, I was fortunate to break bread with many of the Nimnicht family members at The Brick in Avondale. Edward II and Billie III were there along with Ed's wife, Patty; Billie's wife, Amanda; and Lauren Dozier, the marketing director for Nimnicht family of dealerships. It was great to talk about those beginning days.

Of course, no one in Jacksonville could have foreseen that the Japanese would bomb Pearl Harbor in December 1941, the same year Riverside Chevrolet opened. Nor could anyone have foreseen that the United States would be drawn into a great world war. It was hardly a good time to start

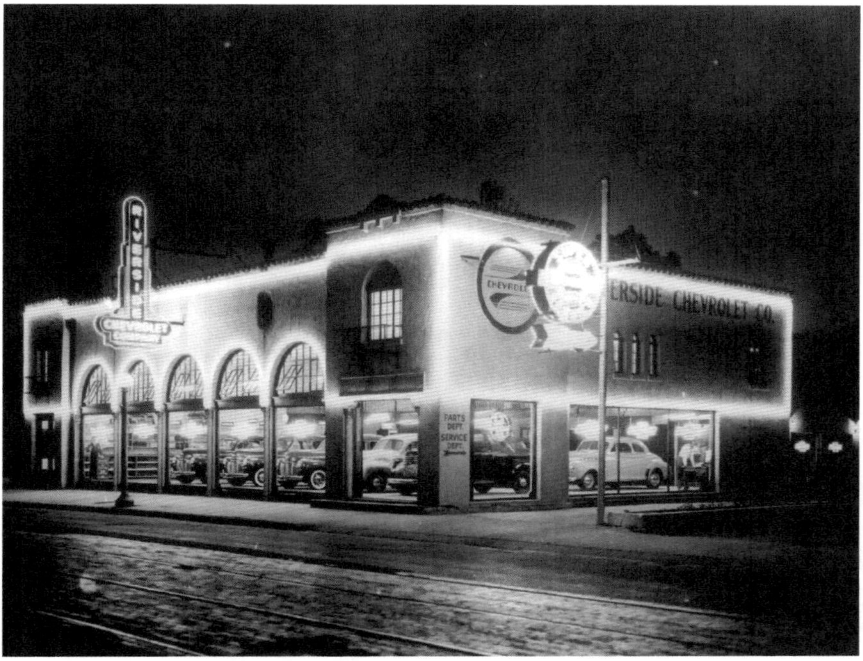

Riverside Chevrolet at night, 1941. *Courtesy of the Nimnicht family.*

any business, much less a car dealership, when manufacture of cars all but stopped and the war machine began preparations for war. It would be especially hard to maintain a car since products that required rubber (like tires) and gasoline were soon rationed.

According to Ed Nimnicht II, Riverside Chevrolet persevered through the war years, and his father, Bill, was able to establish a respected enterprise by providing automotive services and repairs to Jacksonville car owners. "We did a lot of repair work and sold used cars, at least until that war was over, but by 1946, when new Chevrolets were being produced again, Riverside Chevrolet really began to grow."

In 1964, Bill Jr. joined the company as a full-time assistant sales manager, and in 1966, Edward II came into the business as assistant service manager. "Our father insisted that we all learn all aspects of the business, so we worked in every department," Edward II said. He even worked in the machine shop and body shop. He also had a great group of wonderful stories.

"When Hurricane Dora came, we had to move all the cars from the lower-level parking lot to higher ground. The business was located at

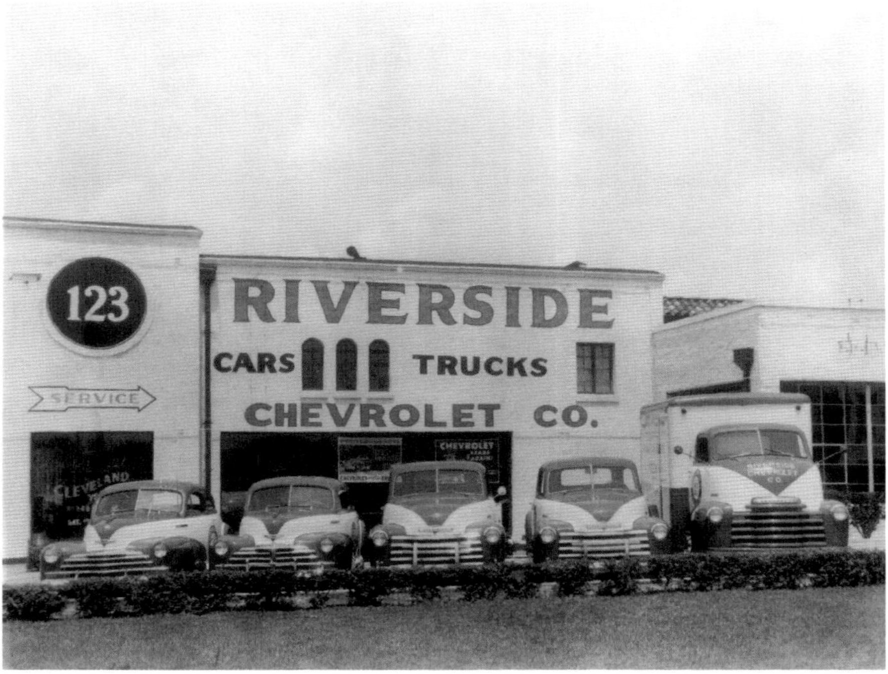

Riverside Chevrolet during the daytime. *Courtesy of the Nimnicht family.*

the place where the Haskell Building is now, so there was a kind of bluff there. The water came over the bulkhead and up so high that it completely flooded that lower level. There was a six-foot chain link fence there with a one-foot barbed wire section on the top. Only the barbed wire was visible after the storm rolled through, and we only lost a few lifts."

Another memory that the family shared with me was "Show Day." That was the day when the newest models were finally revealed to the public, and it was a really big deal back then. It is not at all like it is now, when anyone can see the new models online long before they arrive in the showrooms. No, back in the '40s, '50s and '60s, the display windows were draped so that no one could see in while they set up and prepared the new cars. The employees were to get new dresses, and the dealership provided the female employees with corsages. Flowers were everywhere in the showroom.

At this point, Patty Nimnicht mentioned how fun the National Automobile Dealership Association (NADA) conventions were. "Dinah Shore was sure to come out and sing 'See the USA in Your Chevrolet,' and they also played

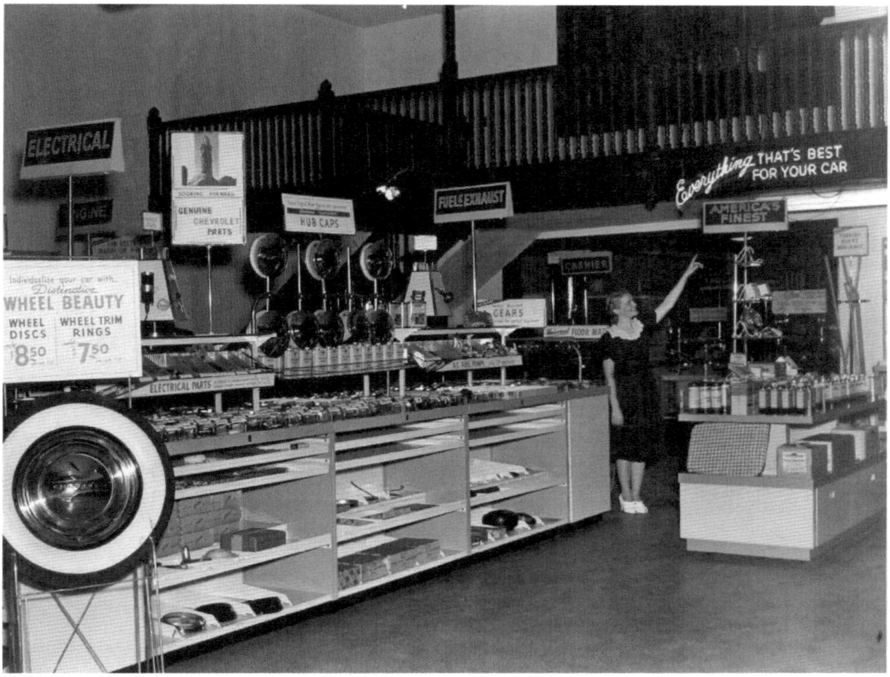

Accessories shop at Riverside Chevrolet. *Courtesy of the Nimnicht family.*

all sorts of American patriotic music which made Mama Nimnicht cry. She loved going to those conventions."

In the late '60s, the dealership was renamed Big R in anticipation of a move from Riverside, and in 1971, the dealership left its original location and moved to 1550 Cassett Avenue.

I met with Lee Nimnicht some weeks later at a Starbucks on Blanding, where we talked cars and drank coffee. Lee really helped me put the history of his family together so that I could better understand the intricacies of the business. First, there was the grandfather Edward Nimnicht, who worked for General Motors in St. Louis. His son, Bill Sr., opened Riverside Chevrolet in 1941 and was the youngest dealership owner in the United States at the time. His sons were Bill Jr. and Edward II, and they continued the tradition by joining the dealership in their teens as part-time workers in all departments, learning the business from the ground up. The members of the next generation to be part of the family business are Billie III and Lee. Their father was Bill Jr. The lineage can be somewhat difficult to straighten out at first glance.

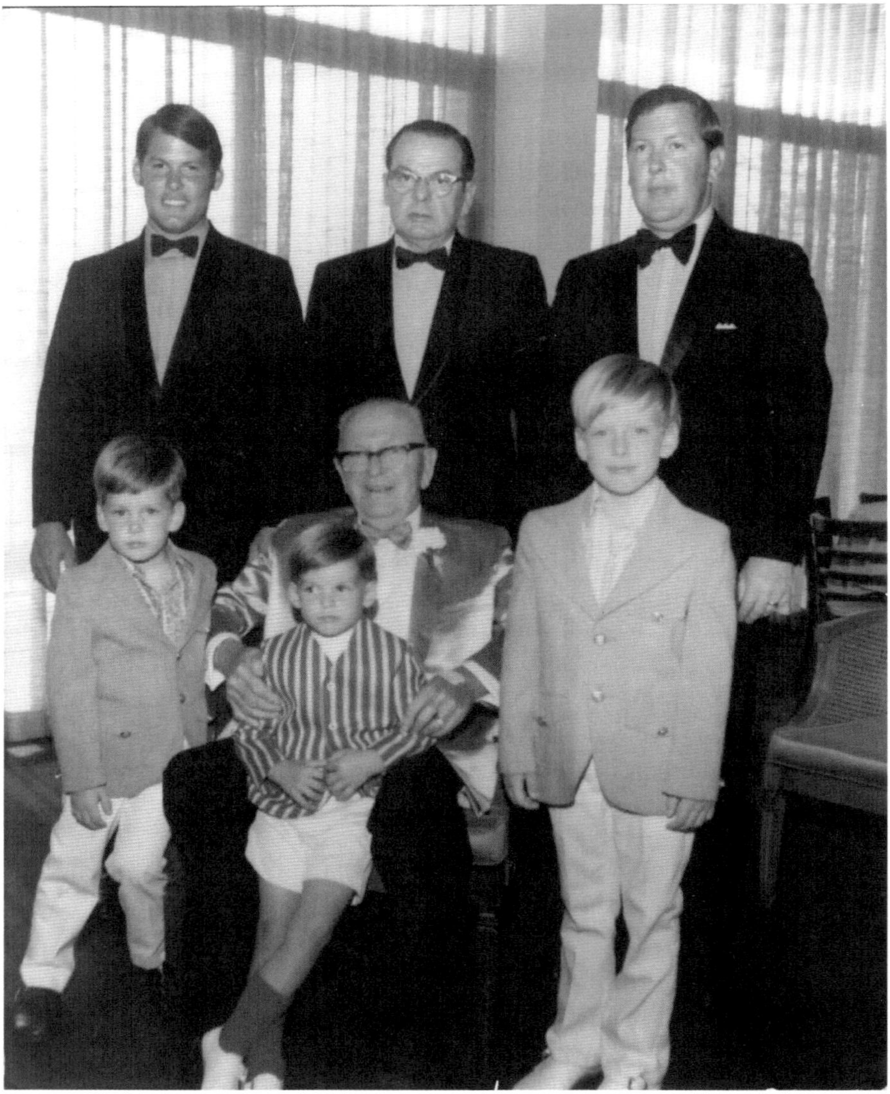

Four generations of the Nimnicht family. *Standing, left to right*: Edward II, Bill Sr. and Bill Jr. *Bottom, left to right*: Lee, Edward Sr., Billy III and Edward III. *Courtesy of the Nimnicht family.*

"In 1993, the family bought the Cadillac Store [Tutwiler's] on Blanding Boulevard. In 2002, we acquired the Hummer franchise at Cassett, and in 2005, Billie and I also bought the Pontiac Store at Philips and Emerson and the GMC Truck franchise from Woolverton Olds on Blanding. When the

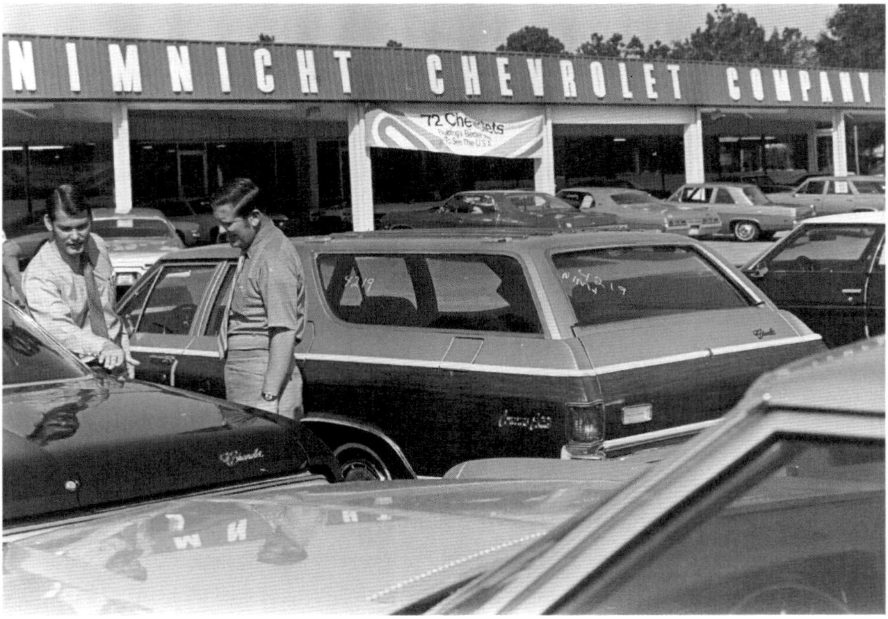

Nimnicht Chevrolet on Cassat Avenue. *Courtesy of the Nimnicht family.*

downturn came in 2009–10 and General Motors went bankrupt, we lost the Hummer and the Pontiac franchises but were able to keep the GMC Trucks. In 2012, we acquired Key Buick, which we split into two parts. One part was Key Buick, Hyundai and GMC and Buick. The other was GMC Trucks. To date, Nimnicht Chevrolet is on Cassett Avenue, and Buick and GMC are on Southside. I work at both locations."

After getting to know the Nimnicht family, it was clear to see that there is a passionate commitment to the franchises they have nurtured and maintained and the community they have embraced for so many generations. They are big supporters of Jacksonville University and University of North Florida. They also support the United Way and the Players Championship, where Lee and Billie's mother was the first female chairman of the event. They also support The Players, the Donna 5K Run and the Gator Bowl Association.

The Nimnicht family of dealerships has always found ways to prosper and grow and contribute to the community. Lee Nimnicht said it well: "Whether it's good times or bad, everything has pros and cons. Since our name is attached to the business, we want to minimize the bad, and we hope that over the seventy-five years we have been in business it's been more positive."

LANGLEY OLDSMOBILE

I bought my first car on my lunch hour at the Teachers Credit Union sale. I was terrified, but I had done my homework and knew the car and the right price. I ran, I did the deal, and went back to teach afternoon classes. When I went back to give them my old car, I couldn't remember the color of the new one. The salesman was appalled.

—*Victoria Register-Freeman*

It is always interesting to discover how mighty things begin, and it is fascinating to me that great businesses, ones that have been part of my life's landscape, got their start in some pretty inauspicious ways. I spoke with Ronald Langley recently and learned a great deal about the beginnings of his family's business.

In the early '40s, Emmett Robert Langley Jr. and his wife, Dorothy, brought their family to Jacksonville, Florida, from Atlanta, Georgia. Langley started selling vehicles for the Superior Coach Company on San Marco Boulevard. Then, in 1952, he was able to procure an Oldsmobile dealership. He then established the dealership at Philips Highway not far from Emerson, which at the time was almost at the end of the world.

In one half of the showroom of his business, Langley Jr. displayed cars. In the other half, he displayed school buses, ambulances and hearses. Langley phased out the coaches in favor of Oldsmobiles, and in what I would consider a very generous act, he took in a tenant, Gordon Thompson, who had procured a Chevrolet dealership and needed the space until he could build his own building. So then half the showroom was Oldsmobiles and the other Chevrolets. Eventually, Gordon Thompson moved on to his own place, just north on Philips Highway.

Ronald Langley also told me that all three of the sons of Emmett Robert Langley Jr. became part of the business in some fashion. Ronald was mostly involved in the business during the summer when he was young and not in school. He eventually went into the insurance business.

His eldest brother, Emmett Robert Langley III, had graduated from the Bolles School in 1951 and attended Virginia Military Institute. While still a cadet, he enlisted in the U.S. Army and left VMI to fight in the Korean War as an army ranger. He received a Purple Heart, and after his discharge, he attended and graduated from the General Motors Institute in Flint, Michigan. While in Michigan, he met his wife, Joyce Waller, who was working at GMI. In 1956, they married and then moved to Jacksonville. He then joined the family business.

His next older brother, James Crawford Langley, also worked in the family business. He attended and graduated from the Bolles School and the University of Florida. Then, after four years in the U.S. Army Air Corps as a first lieutenant, flight instructor, he came back to Jacksonville from Fort Benning, Georgia, and worked full time at Langley Oldsmobile.

Eventually, the Langleys bought property at Edgewood and Highway and started a Lloyd dealership, a German car company. This would become a Willys Jeep dealership, and Langley III ran that dealership, eventually phasing out the Lloyds. They then bought the adjacent property across the street and opened a Datsun (now Nissan) dealership, which James ran that until he retired in 1988.

One of the things Ronald Langley remembers clearly about the early days of the business is that there was a large rocket painted on the side of their family's car dealership at the Philips Highway location. As a result, his father, E. Robert Langley Jr., became known as "The Rocket."

He also remembered being a young teen when he went to one of the dealership conventions. "It must have been in 1954 or '55, when I was about fifteen. I clearly remember being impressed with their showing what the future of cars was going to be like. You'd get in your car, the steering wheel would fold in and the car would take over the driving for you. It would somehow ride on a magnetic force field—rather like hovering over a cable in the road—and you could get on and off as you pleased. That was the future, and maybe they were sort of right. It is not that far off from what will happen with these driverless cars they've been talking about lately."

A few days later, I spoke on the telephone with Joyce Langley, the wife of James C. Langley and the sister-in-law of Ron. She repeatedly told me that her family's life in the auto industry was in a "different era":

> *I was a stay-at-home mom with four little boys when my husband was in the business, and it was just an innocent era, a different era. Sure, business was business, but it wasn't all cutthroat like people might think. There were certain things that we wives did together—sometimes dining at the Green Turtle with other dealers and their families or going to the Strickland's Townhouse. Those restaurants were close to a number of other Philip's Highway dealerships, and we were always friendly with each other.... We wives were also included at the Dealer Association conventions, and we were often on the side watching our children at swim meets. It was just a different era.*

She also could remember some very interesting things about "Daddy Bob" (E.R. Langley Jr.) and his time running Langley Oldsmobile: "He had a 1903 Oldsmobile car in the showroom, and when it was Christmas, he had a big tree in there."

As we talked as if we were old friends, she told me of a time when she was about fifteen: "I would fill up my car with fifteen cents worth of gas and all my buddies ["Lee's Ladies," she called them], and we would head out for a day at the beach. Oh, I hadn't thought of that in a long time."

We both agreed that looking back on earlier times is a really fun thing to do, especially as it concerns cars.

GORDON THOMPSON CHEVROLET

> *I bought a 1969 Chevrolet Chevelle from Gordon Thompson Chevrolet at the beach in March 1969. My husband was in the army stationed at Fort Carson, Colorado. We had been married for ten months. I was proud of myself, as I did this all alone.*
>
> —*Kristine Taylor*

I thought it only fitting that Gordon Thompson Chevrolet be mentioned since so many people remembered that dealership and had things to tell me about it. It all began in 1953 at 2600 Philips Highway, and it was owned by Gordon Thompson Sr. When he died in 1964, Gordon Thompson Jr. took over the dealership, but in 1975, Gordon Thompson Jr. was killed in a shooting accident.

His obituary, which ran on April 4, 1975, in the *Florida Times-Union*, listed his many accomplishments. He was an outstanding athlete—all-Southern football player at Landon Senior High—and he later played at the University of Florida. His nickname was "Stumpy." He was president of Gordon Thompson Chevrolet and was the past president of the Jacksonville Dealers Association and the Southside Optimists Club.

In 1970 and 1971, he was president of the Greater Jacksonville Open Golf Tournament. In 1973, he served as the president of the Gator Bowl Association. He was a director of the Arlington Boys Club, the Baptist Home for Children, Memorial Hospital, the American National Bank, the American Beach Boulevard National Bank and the Hidden Hills Country Club.

In an article published on October 18, 2016, and written by Chamber Williams for the *Florida Times-Union*, Jerry Hamm was described as taking over the dealership after the death of Gordon Thompson Jr. in 1975. Hamm had worked there since 1954. His brother David Hamm joined the dealership in 1978, and his brother Mike Hamm joined in 1992 after a career in ministry.

The article went on to say that the Jerry Hamm Chevrolet dealership on Philips Highway had been sold to Orlando-based car dealer Andrew Ferguson, a former owner of the Orange Park Cadillac dealership. The dealership was then renamed Riverside Chevrolet.

BRUNDAGE/BRUMOS

Another Jacksonville automotive pioneer has to be Hubert Brundage. By the end of his life, he had three corporations to his name: Brundage Motors Inc., a VW distributor to sixty-one dealerships in three southeastern states; Brundage Motors Jacksonville, a VW dealership; and Brumos Porsche Car Corporation. Not only has the Brundage name represented quality product throughout its existence, but it has also been very much a part of the racing scene and has represented the Jacksonville area most respectably in many races.

An article written by Phil Carney for *Excellence: The Magazine About Porsches* and published in February 2004 was provided to me by Jan Brundage, the son of Hubert Brundage, and he assured me that it was most accurate. He didn't mention that he was one of the Brumos drivers early on, though. Once I learned that fact, I quickly e-mailed Jan Brundage and asked him to tell me about his racing career. He wrote back saying:

> *I won my first in class hillclimb in Colorado with my Simca coupe at age 18. Drove in a few local Florida and South Carolina racing events subsequently in Volkswagens, Porsches and Alfa Romeo. My two highlights were driving the first Porsche Carrera Speedster outside of Germany at the Nassau Trophy Races in 1955 and co-driving a Porsche Spyder at Sebring finishing 14th in 1956.*

The article noted that after World War II, Hubert Brundage was running one of three family hardware stores in Miami, and on the side, he was buying war surplus planes and refurbishing them for wealthy Cubans. It was during this time that Brundage became interested in foreign cars. He first bought a four-door Renault for the family, but it was the purchase of a Volkswagen Beetle that set him on a new course in his life. According to the article, "Despite its 26 horse-powered engine, crash box transmission, and mechanical brakes," Brundage thought the VW was a "blast to drive."

Since Brundage's father, Ira Brundage, had raced Ford-Duesenbergs in the 1940s, it was only natural that Hubert would be interested in racing as well. So, he modified a stock VW with a streamlined aluminum skin and entered in the 1952 Sebring Endurance Race. To everyone's surprise, even Brundage himself, he and his brother, Jack, came in eleventh place:

> *The Brundages were no doubt pleased with the Sebring results, but Hubert had an even better idea. A week before Sebring, he inspected Max Hoffman's Glockler-Porsche racing at Vero Beach. What he saw was a tube-frame chassis with a mid-engine installation. So, Hubert designed and fabricated a new tube-frame chassis for his VW Special, reinstalled the car's engine with a special Shorrock super charger, added hydraulic brakes and racing seats and modified roadster body as required.*

Even though the engine eventually "let go," this new design helped Hubert to "outdrag" Bill Spears, who was driving a Ferrari 340 Mexico during a practice run in Tampa. Brundage then teamed up with Glenn H. Curtiss, and they open what they called a "hobby-shop of imported cars." Brundage even displayed a VW sedan at his hardware shop. "After driving and racing the dickens out of VWs," as the article noted, Brundage was sold on the future of VWs in the United States and considered buying a VW distributorship; he began to actively petition for one. The process was long and didn't seem to be going anywhere very quickly.

Then, quite unexpectedly, as Brundage and Curtiss were in the middle of preparing an MG Special for a race at Sebring in 1953, an answer came. They almost lost their chance at having a distributorship for which they had been petitioning. A Mr. Van de Kamp of the Volkswagen headquarters showed up at a race to talk business, and Brundage took exception at the untimely nature of the interruption. He apparently was rather abrupt when he told Mr. Van de Kamp to make an appointment for next week. Brundage said many years later:

> *Suppose he had gotten mad at me for being somewhat rude to him and had given the distributorship to someone else! A lifetime fortune would have*

been lost. My partner and I met with Mr. Van de Kamp the following week and ended up signing up with him for the entire state of Florida as our territory and an order of seventy-five cars to be delivered at the rate of twenty-five cars a month for the following three months. My partner and I had quite a party that night worrying about how we were going to get rid of so many cars.

Hubert and his wife, Nadine, packed up the VW sedan and took two weeks to sign up thirty dealers. Despite its rocky beginnings, this enterprise was off to an auspicious conclusion. Brundage Motors Inc. was basically born in Miami Springs, Florida, and BMI, the Volkswagen distributor, was relocated to Jacksonville in 1957. Early on in 1954, BMI had six employees, and its dealers had sold 102 cars. In ten years, there would be sixty dealers in three states—Florida, Georgia and South Carolina—and 25,000 new VWs sold the year of 1964.

One thing that Brundage also was very involved in was the development of Formula Junior racing and Formula Vee. Formula Junior was designed to allow beginners to get into racing, but as time went on, some racers' cars became overdeveloped and pricey. Brundage, by selling VWs, realized that these cars' quality improved as the base price remained within the realm of the middle-class consumer. Perhaps this is why he designed a new kind of Formula Junior competitor. His idea was to design a single-seater, open-wheeled car with a standard VW engine, transmission, brakes, wheels, tires and suspension.

In 1958, Hubert went to Germany to get a VW to take to Italy, where Enrico Nardi was to construct an aluminum body over the VW base and race car similar to a Formula Junior. Nine months later, the car arrived in the United States. This "prototype" would be sold to George Smith and Bill Duckworth of Orlando for the sum of one dollar. They would use it to duplicate the body in fiberglass and then sell kits for racers to assemble. Their company was called Formcar, and the racing class was to be called Formula Vee.

According to the article, Brundage became interested in Porsches when they began racing at Sebring in 1954. He began petitioning the Porsche KG for a distributorship as early as 1955, but at the time, Maximilian E. Hoffman, a powerful importer of Porsches, prevented it. It would not be until 1959 that Hoffman's dominance over the Porsche market came to an end and Porsche of America Corporation assigned six distributorships. Brundage was one of those. His new company was called Brumos Porsche

Car Corporation (combining the cable address "BRUndage" "MOtors" to make "Brumos").

One interesting thing happened in 1957. Volkswagen banned the racing of its cars because, as the article states, "it didn't want its cars associated with motorsports." Even owners of dealerships were prohibited from racing. So, from 1957 to 1959, no Brundage cars appeared at Sebring or any other major events.

Thankfully, Porsche did not have such a ban, and in 1960, the Brumos race car made an appearance at Sebring. After a variety of mishaps, the first Brumos race car, driven by Bob Holbert Sr. and Roy Schecter, finished second overall.

In 1961, Holbert returned with Roger Penske, and they came in first in class and sixth overall. In 1962, the Brumos car was a 356 coupe driven by Lin Coleman and Pat Corrigan and placed twenty-fourth out of sixty-five. In 1963, Brumos's GT, driven by Lin and Charlie Kolb, posted a DNF, but Brumos had made itself known at Sebring for certain.

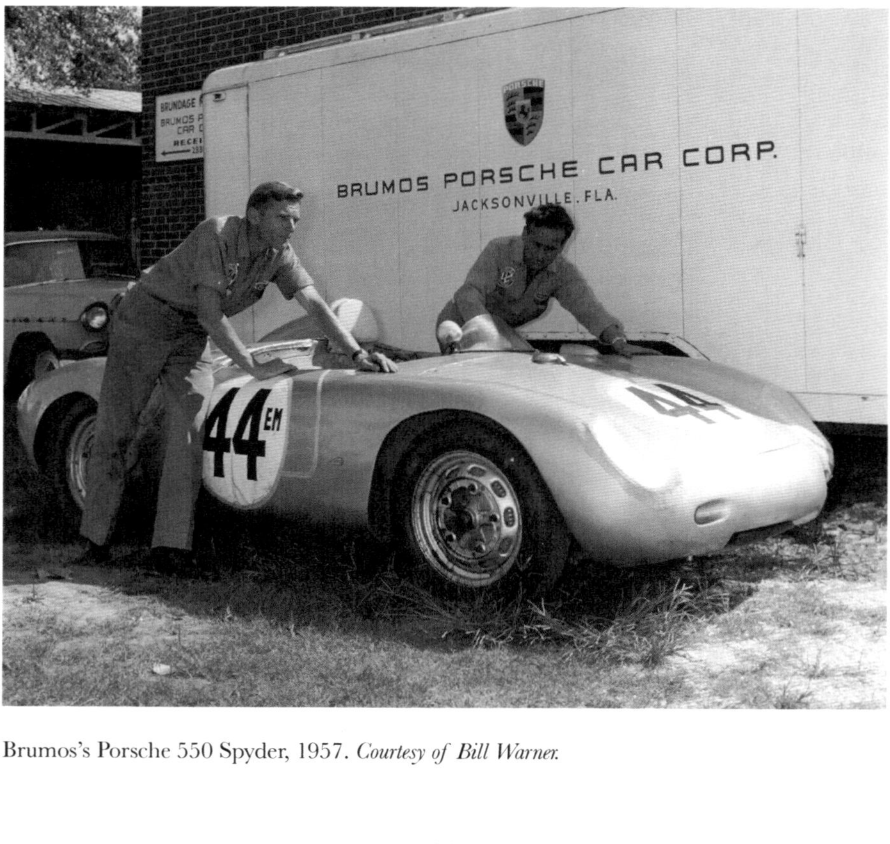

Brumos's Porsche 550 Spyder, 1957. *Courtesy of Bill Warner.*

Sadly, on November 9, 1964, Hubert Brundage died from injuries he sustained when his motorcycle went out of control in front of his home in Mandarin. In an employee newsletter called the *Brumos Bugle*, there was a moving tribute paid to the boss: "Whenever we hear the busy sound of a VW motor up a Jacksonville Street or see a sleek Porsche speeding along a Florida Highway, we may pause and know that his spirit lives on, and will continue to guide us in our daily endeavors. So, Goodbye Mr. 'B' and thanks for allowing us to know you."

There was considerable concern in Germany as to who would run the business now that Brundage had died. His Brumos Porsche Car Corporation had thirty-four dealers in seven states and had become quite an extension to Porsche KG. At Porsche's request, John Von Neumann, the West Coast distributor at the time, was prepped to take over the responsibilities of Brumos. He really wasn't interested in the day-to-day activities of the East Coast business, so when a young Peter Gregg asked him in 1965 if he would sell him the dealership portion of Brumos Porsche Car Corporation, Von Neumann quickly reached an agreement with him. Gregg then approached Nadine Brundage, Hubert's widow, for the rights to the Brumos name, and she agreed.

Von Neumann remained the head of Brumos's distributorship segment until 1969, when Porsche+Audi formed a division of VW of America to handle Porsche importation, distribution and servicing in the United States. Peter Gregg would go on to make Brumos Porsche a dominant force in racing.

After Gregg's death in 1980, his widow, Deborah, assumed ownership and kept Brumos Racing in operation. In about a decade, Bob Snodgrass and Dan Davis bought the business, and it operated until it was sold to the Field Automotive Group in 2015.

CAR MISCELLANY

My dad died when I was in ninth grade, and my mother kept her Chevy II for me to have when I got old enough to drive. My sixteenth birthday was on a Sunday in August, and I had to wait to get my license on the following Monday. I think my mother was more excited about my getting my license so she didn't have to leave work to get me places. She pushed me to learn to drive! I think there was a stop sign on the corner of Hernando and Ponce de Leon that, on my first day out behind the wheel, I nearly ran. She yelled, "Stop!" My mother never raised her voice, but she sure did that day!

When school began in September, I took my neighbors to school…the same kids I had carpooled with to DuPont. I charged them five cents each way, or fifty cents a week, but I did give them door-to-door service. The approximately two dollars I made each week helped defray the cost of gas…that was something like twenty-five cents a gallon. I think I could fill her up for four dollars!

—Etta Fialkow

It is one thing to buy a car. It is quite another to keep it running. An owner must be prepared to care for the purchase—this new car that is now an integral part of life. There is an engine to maintain, tires to rotate and fuel to make it run at all. These needs have all been addressed by the ancillary businesses that grew out of the car industry. I found two such businesses that struck me as being very important parts of the car experience—one that still operates and another that has a Verizon store on the property.

The first business, still in operation, came to my attention when I was researching a different book. For as long as I can remember, there has been an imposing building at the northeast corner of University Boulevard and Philips Highway. In my earlier years, it was the building in front of the screen of the Southside Drive-in, the closest drive-in theater to my house and a place where my teenage friends and I spent a lot of time. Later, when I was doing research for one of a number of projects, I visited the place called Big Chief Tire Company. It seemed to me that the building was a lot fancier than the run-of-the-mill tire store. With its glass/brick window and curved wall, it gave off an Art Deco vibe, although it is really pretty utilitarian-looking. What I found out about it was interesting though.

I was directed to the present-day owner of Big Chief Tire Stores, whom I contacted soon after. Fritz Parman III graciously granted me a phone interview about his family's business. Big Chief was founded by Fritz Parman Sr. on August 16, 1961. He had been born in Corbin, Kentucky, and if his "salesman story" is true, he actually worked for Colonel Sanders at a gas station Sanders owned before the Colonel struck it rich with Kentucky Fried Chicken.

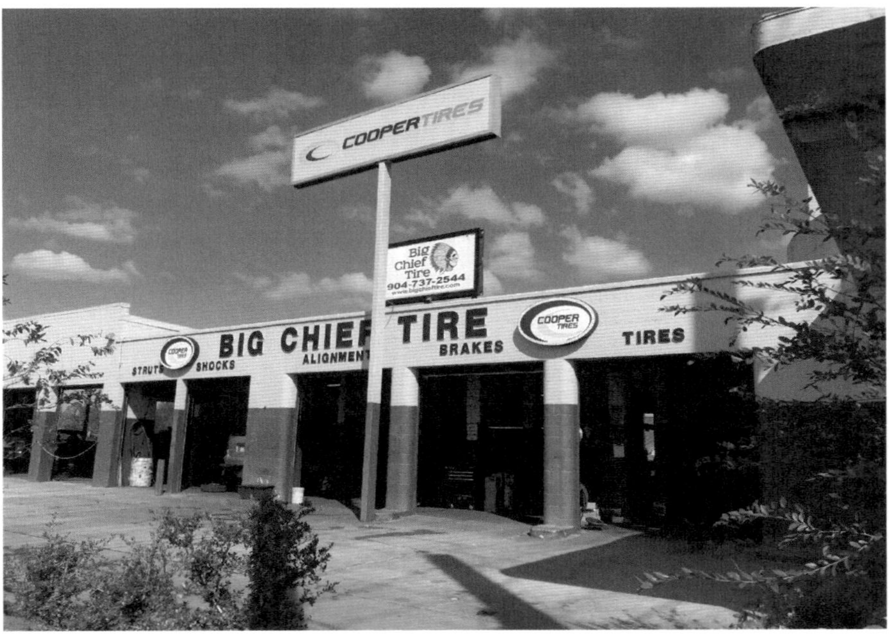

Big Chief Tire at Philips Highway. *Author's collection.*

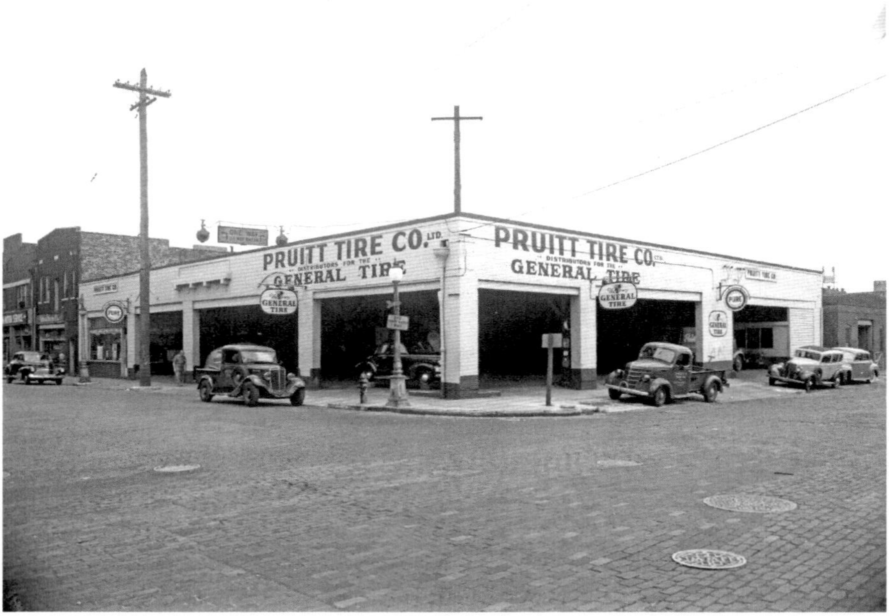

Pruitt Tire Company, circa 1940s. *Courtesy of Florida State Photographic Archives.*

"My grandfather was seventeen," Parman told me, "when he lied and said he was eighteen so he could get into the navy. He worked hard and made it up the ranks to chief before he left the service in 1957. He then tried his hand at being an insurance salesman, selling policies door to door, and he really hated it, but he learned that he was a master salesman. He soon joined Webb Pruitt Tire Company and quickly excelled there."

Parman went on to say that in 1961, his grandfather bought the two Pruitt's Tires Stores and changed the names to Big Chief, since he was indeed the "big chief" of operations as well as having been a chief in the navy. He opened the two locations—one at Normandy Boulevard and the other on Roosevelt Boulevard—under the new name. In the late '60s, he opened a Blanding Boulevard location, and in the mid-'70s, Parman Sr. opened the University Boulevard and Philips Highway location, as well as a store at the corner of Lem Turner and Dunn Avenue.

Big Chief was a true family enterprise. Fritz Parman Jr. came on board when the stores opened and became a driving force in the business. His mother, Dorothy Williams Parman, was the main administrator of the business. When Fritz Parman III, part of the third generation of Parmans, was a junior in high school, he started working at the family business. He worked with his grandmother, learning all of the administrative aspects of the business. "When my grandmother died in 1991," he said, "I went through quite a trial by fire—first I lost my grandmother and second, I was thrust into a position to fill her shoes—taking over my grandmother's position."

He survived this set of circumstances well and became the main administrator of the business. When his father died in 1998, the business passed on to Fritz Parman III, and to this day, he is now "big chief" of it all.

Another business that supported the automotive industry was the one that provided cars with fuel. They were called "service stations." Such places were advertised as happy places where men wearing Texaco Stars would come out of a building singing and dancing their way over to service your vehicle—something that might give a modern person the creeps. Other ads offered to put a "Tiger in Your Tank," another bizarre possibility. In any case, these places were not only where a person could buy the fuel needed for the car, but they were also where attendants would check your oil level, wash your windshield and pump the gas—all this while you visited the restroom or the store within for colas, Lance crackers and any of an assortment of candies.

One such service station took up a corner at the intersection of University and Old St. Augustine Road (although, to my knowledge, there was no singing and dancing there). It was an Atlantic Richfield Station, which would eventually become an ARCO Station and then a Mobil Oil station. It was owned and operated by a first-rate mechanic by the name of Ed Andrews. He was the proprietor.

On a brisk October afternoon, I had the opportunity to interview his wife, Joan, and his daughter, Brenda J. Baker, who brought her husband, Larry. We met at the Starbucks on San Jose and St. Augustine Road and talked an hour away over cups of coffee.

"We came to Jacksonville in 1963, the same year that Kennedy was assassinated," Joan Andrews said. Her husband "went to work with his older brothers, Andy and Cy, at the Sinclair service station, which was where the Hooters on Goodbys Lake is now. They kept calling saying that Jacksonville didn't have any good mechanics, that they needed mechanics down here badly. So we came down, but after a while, the brothers didn't get along and they began to argue. Probably over money. So, Ed said, 'That's it! I'm going to get my own place.'"

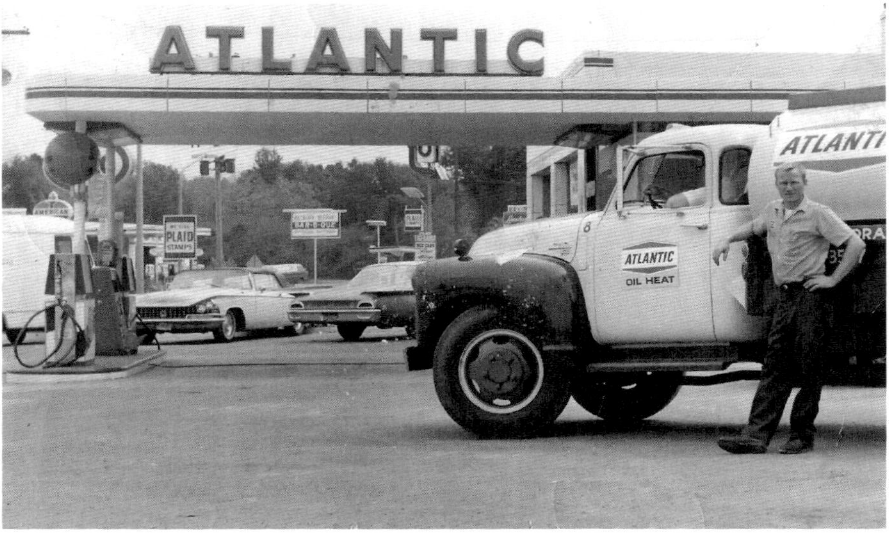

Edward Andrews in front of his service station. *Courtesy of Brenda Andrews Baker.*

After that announcement, he soon became the proprietor of the Atlantic Station on University and St. Augustine. It was available, and he really had to wheel and deal to get it. Brenda was just a little girl when they arrived in Jacksonville. She attended Assumption Catholic School, and then when the family moved closer to her father's work, she attended San Jose Catholic School. Eventually, she would attend Samuel W. Wolfson Senior High and graduate in 1974. From there, she went to the University of Florida and majored in health science. She was the first in her family to graduate college. She's been a teacher, a pharmaceutical sales representative and a life insurance agent. She presently lives in Georgia.

Baker has many wonderful memories of her father and his days at the station. When her father got his own business, he eventually put her to work. While she was a preteen, Andrews let her earn money by cleaning the bathrooms, helping with the Lance and Coca-Cola deliveries and running the credit cards for him while he worked in the garage. She occasionally pumped gas, but when she became a "big girl," she didn't want boys to see her working there. That is when she stopped working at the station.

"I do remember very fondly," she said, "how on Thursday, we Andrews kids and our mother would get in our old woody station wagon to go shopping. First we would have to go to the station to get money from my father. He always had a big roll of money in his pocket since so many people paid in cash in those days. I really thought we were very rich. We would then go shopping at the A&P in Lakewood, go have lunch at the McDonald's at Beach and then we'd go to the Mandarin Supermarket before we went home."

Larry Baker repeated a great story that Brian Andrews, his brother-in-law, told him the night before:

> *One of Ed's customers came in the shop with a brand-new alternator and asked Ed to put it in his car. Ed put it in and said, "That will be six dollars." The customer gets really mad and says, "Six dollars? I could have done it myself for six dollars." Some time goes by, and he comes in with a brand-new tune-up kit and he asks Ed to use it on his car. Ed does, asks for six dollars, and the guy gets angry again. Not too long after, Ed took some raw bacon and several raw eggs one morning to the Derby House on his way to the station. This is where the angry guy worked as a cook. Ed says to him, "Here, cook these for me for twenty-five cents." And the angry guy says, "I can't do that," and Ed says, "Why not? You are always asking me to do things like that." Ed never had any problem with this guy ever again.*

After we all laughed, Brenda remembered what it was like back then. "During those days, my family only drove fixer-uppers, like the woody, and I recall an old Ford Fairlane we had that had those weird, round tail lights. I also remember being teased by the kids at San Jose Catholic School about our dad being a 'grease monkey.' It really bothered my brother, but I just told them to shut up."

In the grand scheme of things, however, those taunts melted away when Ed Andrews died in 2010. The St. Joseph's Catholic Church in Mandarin was filled to overflowing at his funeral, with his many friends and people he had helped. Baker was overwhelmed with gratitude.

"It shouldn't have surprised me. My father was a very kind man. He had so many friends, and he helped a lot of people. He was known to take bartered goods when customers were running low on cash. And he did all the work on the cars of the priests and nuns of the San Jose Catholic Church, for family friends and for well-known locals. And my father worked hard—every day except Sunday—from 8:00 a.m. until it got dark."

Eventually, Ed Andrews left the ARCO on University. Baker showed me an article from the *Jacksonville Journal* that showed the reason behind the move. The headline read "10 Years to Build Business; 10 Days to Pack and Get Out." Mobil Oil Company, which had bought out ARCO's Florida holdings in 1972, had "targeted this unit for disposal because it was not in good repair." It had estimated that it would cost $20,000 to repair the one-man, two-pump station.

Andrews did admit that Mobil had offered him other stations, but these also had financial issues and were, for the most part, too far from his home. Still, his picture is there in the newspaper for all to see. He had put up a sign that read, "After 10 Years Mobil Says Get Lost!" and believe it or not, the Mobil spokesman said that this display would not hurt Andrew's chances of getting another Mobil station if he wanted one. J.R. Moreland, district sales manager for Mobil, said, "He's paid his bills and done a good job for us in that location."

After the ARCO/Mobile station, Ed continued to be a mechanic working with his brother Cy and Cy's son, Bill Andrews, at a three-bay Exxon station on the corner of Beauclerc and San Jose Boulevard. When Bill passed away in his early fifties, Ed could no longer manage the workload, so he got a shop down off Powers Avenue that was basically a large storage unit. Here he serviced only his best customers' cars. The place was hot and had poor ventilation, and his lungs bothered him. He eventually tired of working at the Mobicare that his son operated. By 2001, his health started to fail. He retired and lived quietly with his wife until he died in 2010.

"I am very proud of my father," Baker said as we came to the end of our conversation. After all we had discussed, I certainly could understand why she felt that way.

THERE IS ANOTHER PIECE of Jacksonville automobilia that has fascinated me: fleets of cars. In searching for photographs for this project, I came across several very powerful pictures. The first was of a fleet of Chevy police cars that had been arranged in such a way that they spelled "Duval." This picture was taken in 1959, and the policemen stand at attention in the picture.

At first I thought this was a fleet of Fords used for a promotion for Duval Motors, which I thought had provided the cars. When I looked more carefully, however, I realized that this was indeed a fleet of Chevrolets (based on what I could see of their taillights). I suppose I was understandably confused. I had recently viewed a DVD called *A History of Making Value*, which had been prepared by the Scott/McRae Automotive Group (which can trace its roots back to Duval Motors, a Ford company) to celebrate its eightieth anniversary. In the DVD, Nelson Eason, government sales manager of the Duval Fleet Department, said, "When the city

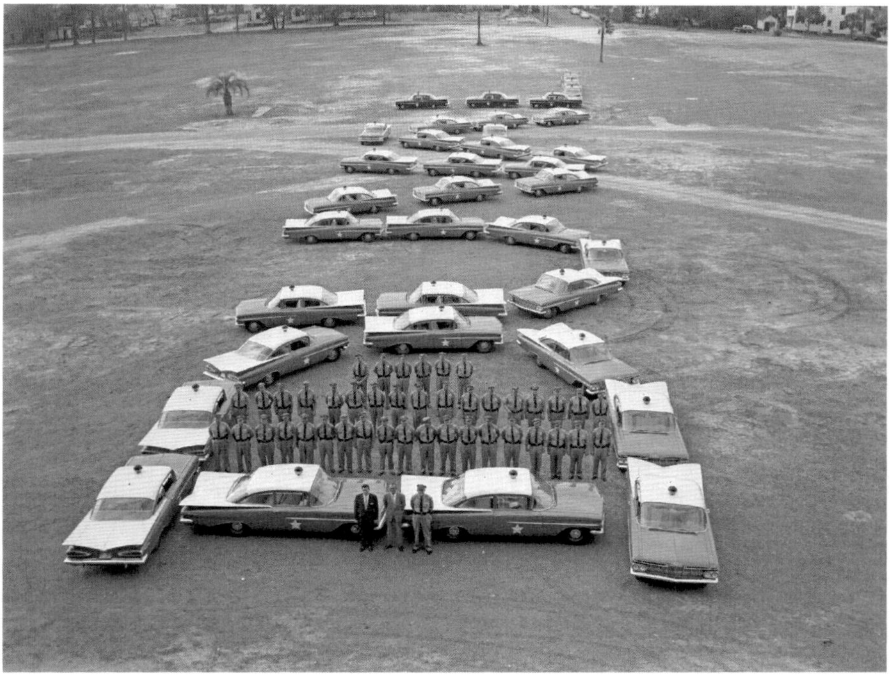

Duval police fleet. *Courtesy of Florida State Photographic Archives.*

[Jacksonville] was sold on the fact that the only kind of police car that they could have was a Chevrolet, Mr. McRae. took that as a personal challenge. We took on the city council, the Sheriff's Department, the Motor Pool and everybody else, and I think he enjoyed the fight more than the fact that we won and we sold them hundreds of cars."

Of course, this picture was taken in 1959, so when McRae sold the police on Ford vehicles, it had to have been in the 1960s. I am also uncertain why this picture was taken in the first place. It may have been advertising, or it may have been to show how well equipped the county police force was. Not only did the cars have to be parked just so to spell "Duval" and all the police force had to be dressed and ready, but the camera also had to be elevated to a certain height so that the whole word could be read. It must have taken hours to prepare the photo shoot. Still, the end product resonated with me.

I also found another great picture of a fleet of cars—limousines, to be exact. These cars looked so spiffy with their uniformed drivers standing at attention next to the driver's door. This picture, taken sometime in the 1940s, shows a number of luxury cars (Chryslers, I think) lined up in front of Imeson Airport—once the main civilian airport of our city. Seeing this picture made me revisit all the times I went there to catch a plane to Dallas, where my grandparents lived, or where I had to say goodbye to friends who moved far away from Jacksonville.

Sandy Strickland said in a "Call Box" article published in the *Florida Times-Union* on August 28, 2017, that the Imeson airport building was built in 1941 and was part of the Work Projects Administration (WPA). This building is certain to be loved by many of us who grew up here. I especially love the building because of its Art Deco elegance. It felt very modern somehow.

Finally, I found a photo of a fleet of armored cars from the Guy Marvin Armored Car Division. This 1949 photograph shows three armored cars and six drivers all lined up in front of the Train Terminal, or what is now called the Prime Osborne Convention Center.

I thought there might be more information about the armored cars, so I searched for this same company, and I was not disappointed. According to the online *Daily Record*'s "Fifty Years Ago This Week" article, it was during the week of March 26, 1962, when the Guy Marvin Armored Car Division suffered a hijacking. As C.E. Griswald and Edgar Strickland were delivering the $88,269.96 receipts from the Jacksonville Kennel Club to the company's location at 122 Jefferson Street, two unmasked gunmen disarmed the driver and the guard and then took them on a wild ride. The driver, who was calling the employees inside to alert them that they were opening the vault, was struck from behind. He was stunned, his gun was taken and he lost his glasses. He was then forced to drive the armored car away. The guard had been bound with adhesive tape and was forced into the back of the car by the other thief.

The armored car went down Jefferson, and when it passed Adams, the gunman sitting beside Griswald honked the horn twice. Immediately, another car pulled up behind them and followed them. At 2091 Run Street, the armored car stopped. Here nine bags of checks, currency and coins were removed from the armored car and placed in the car that had been following. Griswald was ordered to the back of the armored car, where only his hands were bound, since the robbers had run out of tape. The thieves locked the employees in the back and left. It didn't take but a few minutes for them to free themselves. Griswald made it to a telephone and called the police.

Using one fingerprint that was left on the tape, being alerted that someone had just bought a car with cash only and using a lineup where Griswald and Strickland identified the two gunmen, five arrests were

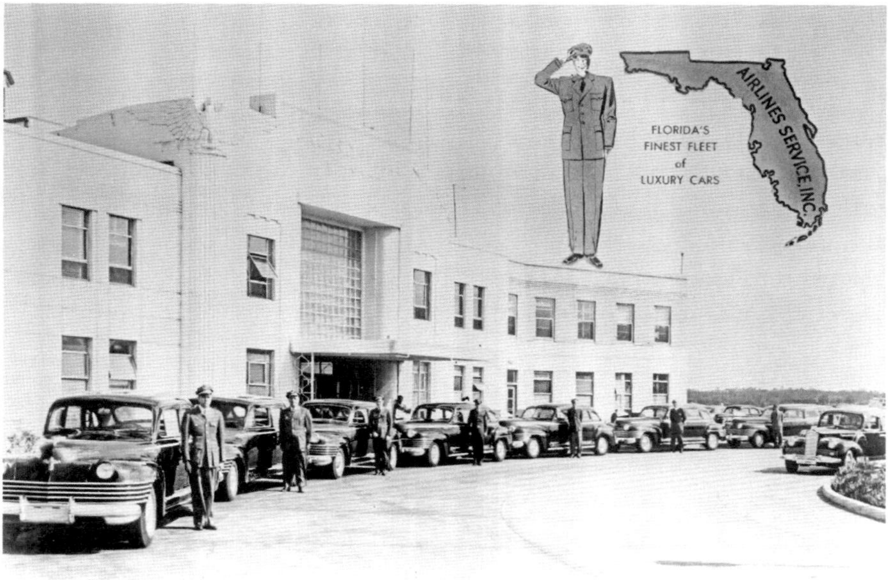

Limousine Fleet at Imeson Airport, circa 1940s. *Courtesy of Florida State Photographic Archives.*

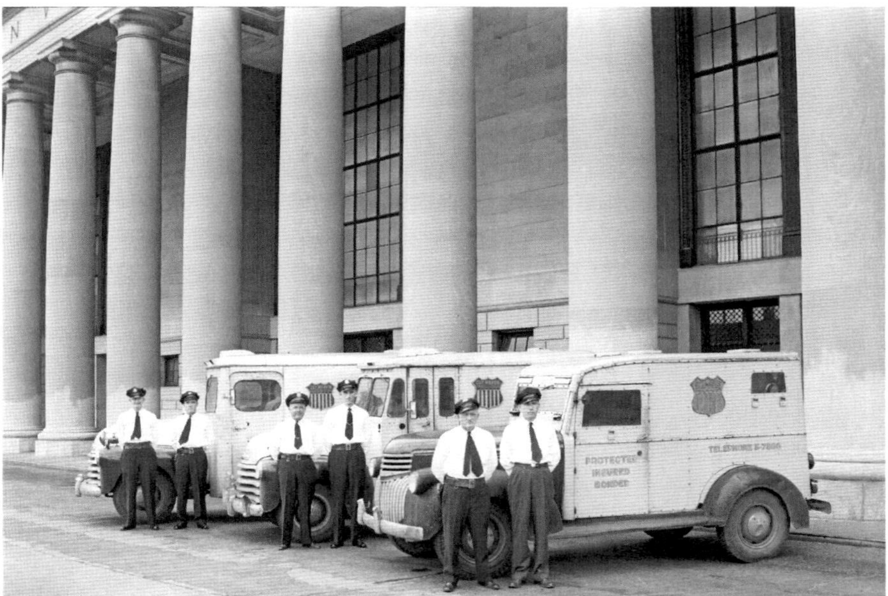

Guy Marvin armored cars in front of train terminal, now the Prime Osborne Convention Center, 1949. *Courtesy of Florida State Photographic Archives.*

eventually made. Sadly, only $46,167.90 was recovered. The remainder was not found.

As I read this news story, I came to the realization once again that the past was not always so sweet and wonderful. The more I research things from the past, the more I know that it was not always rosy. It was all too very real and gritty. Sometimes there were even evildoers lurking in the shadows to take the cash receipts from the dog track. Even so, it is helpful to know that when things seem bleak in the present, looking back can remind us that here and now is actually a good place to be. Just ask Griswald and Strickland.

RACING AND LOCAL DRIVERS

My first car was a 1960 white Impala convertible, and she was awesome! She was about a block long, and originally, she had been my mother's car. When Mom got a new one, I got hers.

So one day, I am driving along in my beautiful car on what was becoming Interstate 95. If you will remember, they were building it one stretch at a time. Anyway, I've got the windows open, and I'm listening to the radio, WAPE, and I remember looking down and realizing the speedometer went all the way up to 120 miles per hour. So, I went, "Hmmm, I wonder…." I opened her up, let her rip and it was awesome! It wasn't long before I ran out of numbers, and pretty quickly I was running out of road. It was then I got to find out how good my brakes were.

Still, that ride was probably one of the most exhilarating things I have ever done. To this day, my mother has no idea that I ever did this, so I can't let her read this book.

I didn't have my Impala very long because I just had to have a brand-new 1970 burgundy Camaro. It was a great car, but I never had the same thrill as I had as when I hit 120 in my Impala.

—Barbara Lewis Milano

The love of racing automobiles in Jacksonville goes back to the earliest times of cars. As soon as two automobile owners got together, it was sure to bring out comparisons, and those comparisons lent themselves very easily to the idea of whose auto could go faster.

Sadly or not, Jacksonville never became as famous for racing as Daytona or Sebring have, but that does not mean that Jacksonville wasn't important in the beginnings of the sport. The first official automobile races here are mentioned in Davis's *History of Jacksonville*, and they were held on April 12–13, 1906, at Atlantic Beach. The overall winner was Joe Lander of Atlanta. Davis said that Lander won in a Thomas car going five miles in 4.55 minutes and one mile in 57.4 seconds, breaking existing world records for stock cars.

However, according to an article written by Bill Warner for the program of *The 13th Annual Amelia Island, Concours d'Elegance in Amelia Island*, March 9, 2008, speed record runs had been going on in Jacksonville before this time. It was as early as 1905 when Claude Nolan himself raced other automobilists on the beach from Jacksonville to Ormond Beach and back. His car proved that the eighty-five-mile trip could be made in under seven hours.

In 1909, Jacksonville was part of the Great Endurance Race from Tampa to Jacksonville and back again. In Tampa, Florida, there is a marker in front of the Tampa Daily Times Building, put there in 1978 by the Tampa Historical Society in cooperation with the AAA Clubs of Florida:

> *GREAT 1909 AUTO RACE*
> *From this point, in front of the* Tampa Daily Times *building on Nov. 23, 1909, was started the Tampa-to-Jacksonville "endurance run" to promote good roads in Florida. Sixteen vehicles, plus the "confetti" pathfinding car and a press car, made the grueling four-day round trip. There were many breakdowns, an accident or two and numerous other tribulations as the early models made their way through the mud and the woods and across streams. The promotion did indeed spark bond issues and construction of highways in the state. The event was sponsored by the Tampa Automobile Club and the* Tampa Daily Times.

There is a fascinating article titled "The Real Winner Was the Beach: From Tampa to Jacksonville" on a website called Something About Everything Racing. The article was originally written by Beverly Rae Kimes for *American Heritage*. In it, Kimes said that the race was the brainchild of courthouse reporter Willis D. Powell of the *Tampa Daily Times*, who had just begun writing a column about the growing automobile industry. A convention was held in Tampa, and the course was set, with many civic leaders clamoring to have the race pass through their towns. "The event was designed to draw attention to the need to improve Florida's roads…most of the state's roads beyond city limits were little more than sandy cattle and wagon trails."

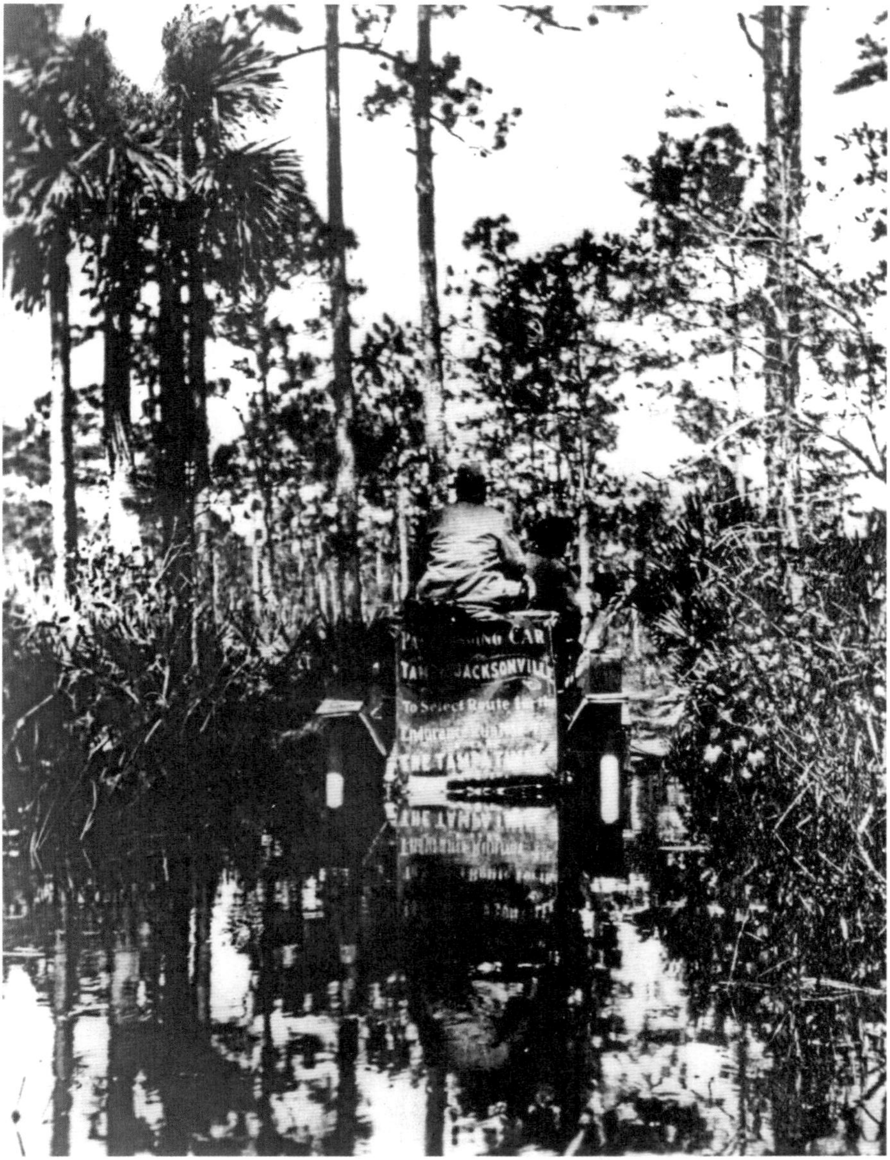

Great Endurance Race of 1909, mired in a swamp. *Courtesy of Florida State Photographic Archives.*

The event was to take four days, and the schedule looked like this:

Day 1: Tampa, Dade City, Brooksville, Ocala and Gainesville.
Day 2: Starke, Jacksonville and St. Augustine.
Day 3: Daytona Beach, DeLand, Orlando and Kissimmee.
Day 4: Bartow, Plant City and Tampa.

At the end of the third day, the drivers made it to the Graystone Hotel in Kissimmee on November 25, 1909, where they had a Thanksgiving banquet. The hotel was full, however, so they had to spend the night in the nearby opera house. The next day, at the end of the race, Horace Williams of St. Petersburg was declared the winner of the 543-mile course. His run was fifty-three hours—just over ten miles per hour. He received a silver cup and a five-acre homestead in nearby Zephyrhills, north of Tampa. The second-place team of Houston Wall and Perry Snow, driving on behalf of an owner, won $100. They were supposedly only fourteen years old.

F.W. Wood, a local Tampa bank president, was among the last to finish. He was, however, undeterred. He said, "The run is over, but it will not be forgotten," and the *Tampa Daily Times* noted of the race, "Improved roads… will attest its influence and no more monumental work could be asked for as the sequel to this labor of love and sportsmanship."

In Jacksonville, there is a marker in front of the old Hayden Burn's Library, which is now the Jesse Ball DuPont Center, a nonprofit complex in downtown Jacksonville. The marker indicates the "Turn Around

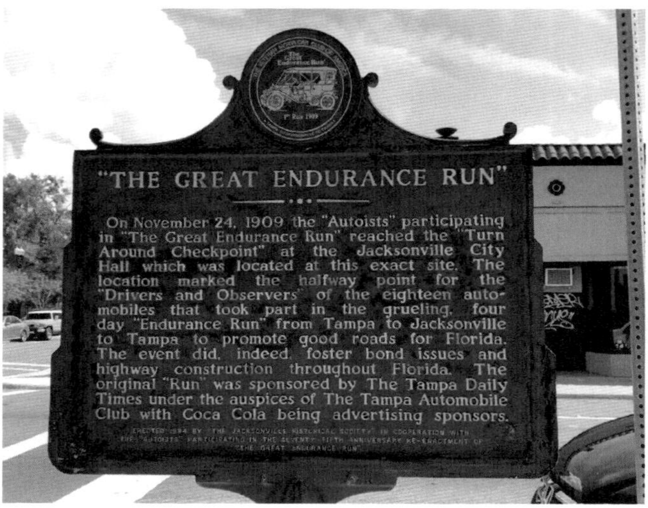

Midpoint of Great Endurance Race of 1909. *Author's collection.*

Checkpoint" at the location of the former Jacksonville City Hall where the "autoists" participating in the "Great Endurance Run" arrived on Day 2 of the race. The marker was erected in 1984 by the Jacksonville Historical Society along with some of the actual "autoists" who had participated in the race to commemorate the seventy-fifth anniversary and to reenact the race.

Barrel or pillar races became all the rage at the beaches all along the North Florida coast—Daytona Beach, Ormond Beach, Ponte Vedra Beach, Pablo Beach/Jacksonville Beach and Atlantic Beach. The courses were defined by barrels or posts that marked the farthest extent of the course.

World speed record attempts also remained very popular as time moved forward. Warner's article said that Bob Burman set a record at Pablo/Jacksonville Beach on March 10, 1911. "Bullet" Bob, as he was called, drove his "Buick Bug" on the 20-mile Atlantic/Pablo Beach course (10 miles down and 10 miles back) in 13 minutes, 11.92 seconds to beat the previous record set by Vincenzo Lancia at Daytona Beach. Later in the same day, ten

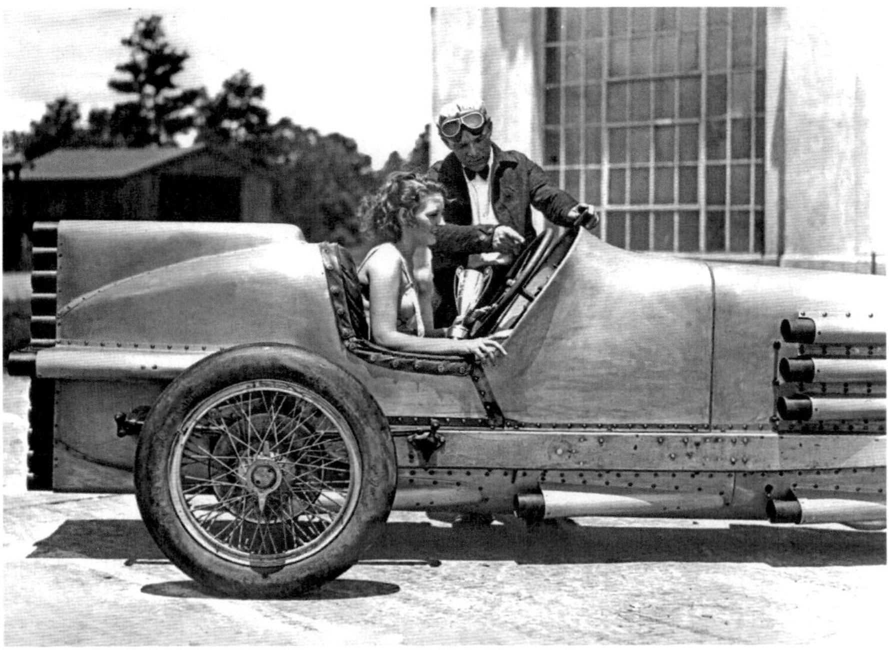

Sig Haugdahl with Mary Pickford, 1930. *Courtesy of Florida State Photographic Archives.*

thousand people watched as Burman drove the "Blitzen Benz" for the fastest mile at 30.25 seconds (119-plus miles per hour).

There were a number of famous racing personalities who raced here in those early times. Sigurd "Sig" Olsen Haugdahl was one example. He was born in Norway, grew up in Minnesota and then came to the South and made quite a name for himself in the annals of racing. He was so famous that there is a publicity picture of Mary Pickford being shown the ropes by Haugdahl as she sits in his Wisconsin Special race car.

According to the article by Warner, Sig Haugdahl was probably "the most successful of all of the Florida based racers" in that early time. In March 1922, he set a world speed record at Pablo Beach going 1 mile in 22.3 seconds. One month later in Daytona, he became the first man to go 3 miles per minute, or 180 miles per hour. This feat he achieved in his rocket-powered Wisconsin Special. This was 26 miles per hour faster than the previous record. Haugdahl was the IMCA champion from 1927 to 1932, and in 1934, he retired from driving. He eventually settled in Jacksonville. When he died, he was buried at Evergreen Cemetery.

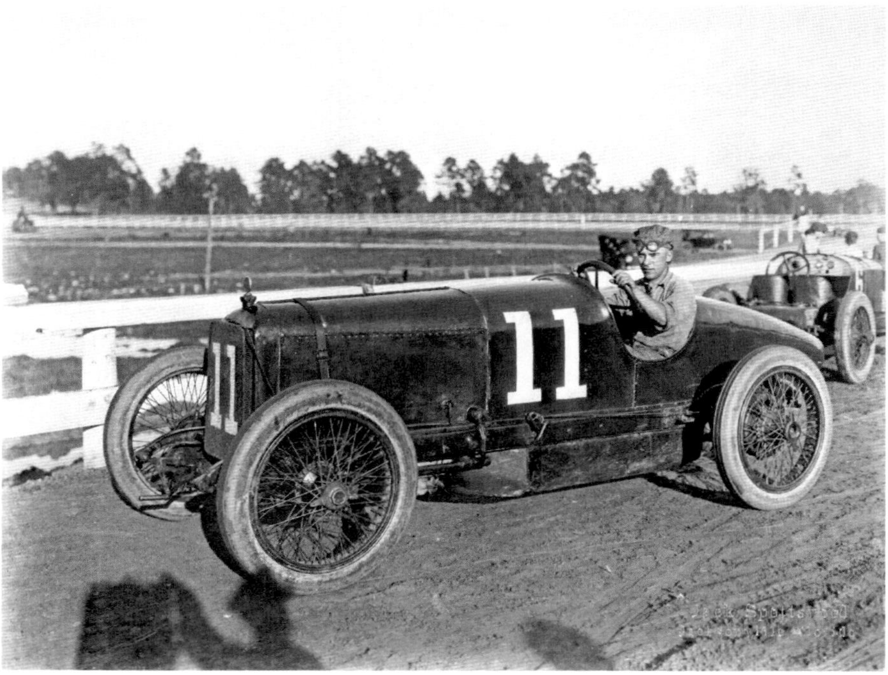

Sig Haugdahl in Jacksonville, 1920. *Courtesy of Florida State Photographic Archives.*

One thing all race fans can be thankful for is that Haugdahl promoted races in Daytona along with Bill France, and although he bowed out, Daytona Beach became a racing mecca. He and France are credited with bringing professional racing to Daytona Beach. Speed record runs ended on Daytona Beach in 1935 because of the dangers of narrow beaches, so such races were then run on the Bonneville Salt Flats.

THERE IS ANOTHER RACER who has some ties to Jacksonville: Bert Deacon Moyer. According to an article in the *Florida Times-Union* written by Dan Scanlan and published on June 13, 2009, Moyers began his racing career around 1923 as a riding mechanic on dirt and beach tracks in Alabama, Georgia and North Florida. Scanlan wrote:

> *He* [Moyers] *raced his Exide Battery Special, then built the Simplex Piston Ring Special using a 1928 Chevrolet engine and brakes, 1926 Chevrolet rear axle and a shapely radiator from a 1927 Whippet, atop a shortened, narrowed 1927 Essex frame.*
>
> *Moyers raced it on courses with turns marked by barrels set a mile to a mile-and-a-half away on Daytona Beach, Pablo Beach (now Jacksonville Beach), Camp Foster (now Jacksonville Naval Air Station) and other sites around North Florida and South Georgia. One photo shows it on Pablo Beach, with Moyers leaning against a back wheel with some road cars, the pier and a Ferris wheel in the background. Another family photo shows Moyers in No. 88 among a row of racers at the 1932 Fourth of July race on Ponte Vedra Beach.*

Scanlan noted that Moyers sold No. 88 in 1935, and in the mid-1940s, his sixteen-year-old son got it back and stored it in his grandmother's Ortega home. Eventually, in 1963, the car was moved to the son's East Arlington home, where he began a restoration project of the car. Illness forced him to stop, and he died in 1995. In 2006, Bill Warner, founder of the Concours d'Elegance, learned that the classic beach racer had been stored in a garage for forty-three years. Warner and Randy Kimberly, a car restoration expert, bought it along with all manner of paperwork and memorabilia saved by the family. Together they restored the car and unveiled it for the Moyers family at the 2007 Amelia Island Concours d'Elegance. It won awards at the Antique Automobile Club of America shows in 2007 and 2008.

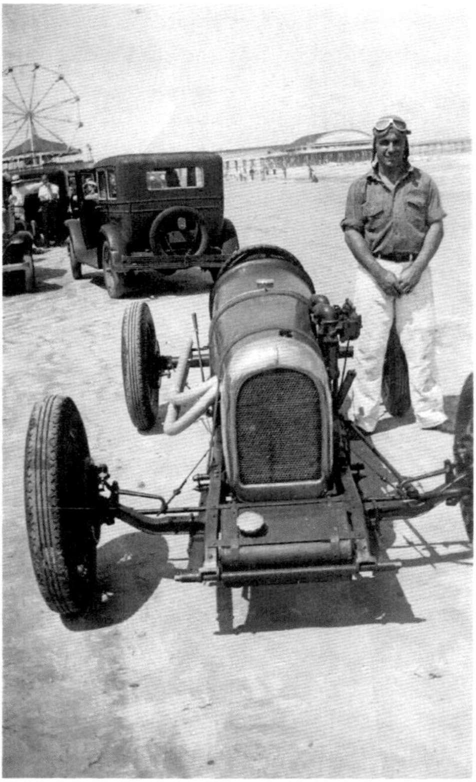

Left: Bert Deacon Moyers on Jacksonville Beach. *Courtesy of Bill Warner.*

Below: Bert Deacon Moyers at Ponte Vedra Beach. *Courtesy of Bill Warner.*

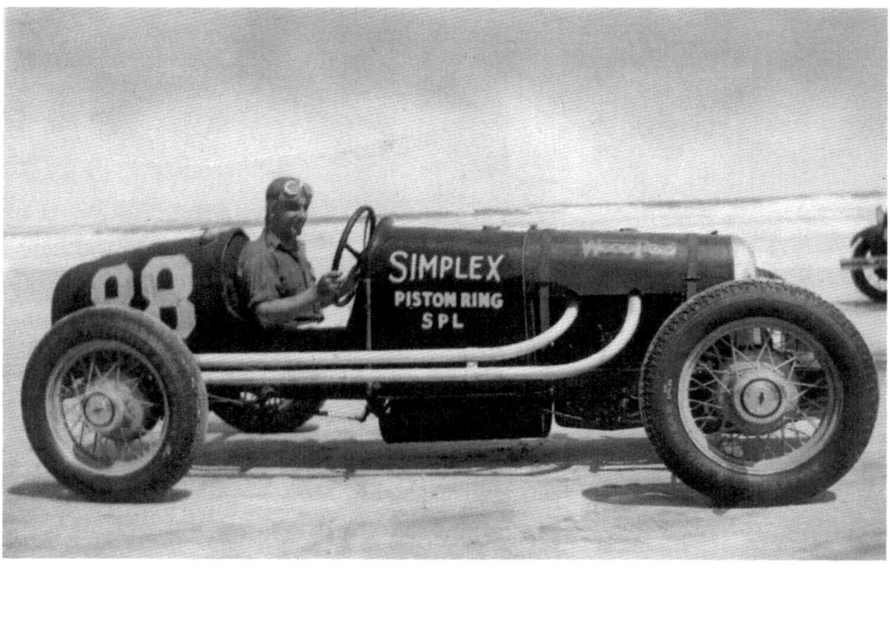

ANOTHER EARLY RACE THAT was staged here in Jacksonville was the Jacksonville One Hundred Miles Auto Race Classic Run, which took place on April 1, 1934. It was sponsored by the Jacksonville Post Florida National Guard, and it took place at Camp Foster, which would become Jacksonville Naval Air Station in 1940.

On April 1, 1934, an article in the *Florida Times-Union* ran with what appeared to be a racing photograph framed with artists' renderings of race drivers around the border to announce the event. The short article beneath noted:

> *Wheel-to-wheel and hub-to-hub dirt track and speedway stars will battle it out for cash and glory at Camp Foster this afternoon. One hundred miles of grueling competition starting at 8 o'clock, over a two mile course with two turn embankments on the end of the course and a 100 foot wide sweeping turn on the other side, three quarters of the course hard surfaced and the remainder with regulation dirt-track surface, many spaces such as the above will be on tap for sports fans when the starter flag sends the field of 38 speed demons on their way. Pictured upper left is Hart Weston, Canadian auto race champion, and lower Pietro Alberti, Italian-American speed star, who are looked upon to figure prominently in today's 100-mile event.*

The next day, on April 2, 1934, the *Florida Times-Union* ran an article that discussed the outcome of the race in an article with the headline "Wild Sam Purvis Takes 100-Mile Automobile Race at Camp Foster."

What I found interesting about the article from a journalistic perspective is that someone died in one of the race cars just before the race, but that sad

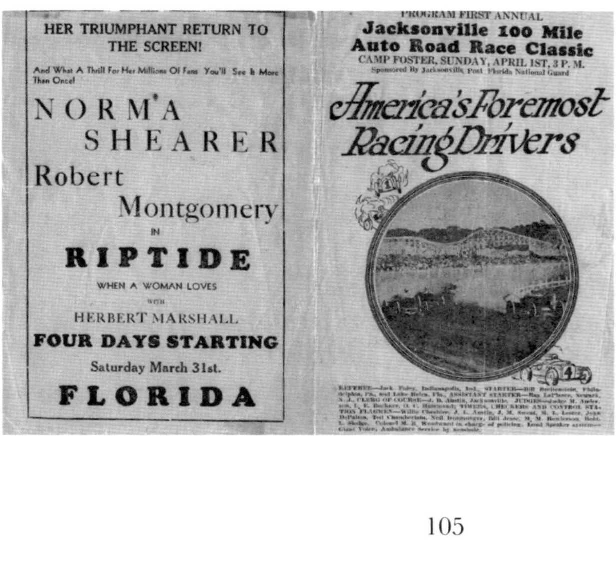

Jacksonville 100 Mile Auto Road Race Classic program. *Courtesy of Bill Warner.*

fact was not mentioned in the lead paragraph. Even so, a second headline read, "Fatal Injuries Were Sustained by W.G. Folk." The article said that W.G. "Bo" Folk, thirty-one, of Jacksonville, had died from "injuries received in a warm up run. Folk who ran down the home-stretch straightaway in his President 8 with tremendous speed, overturned on Turn No. 1, crashed into a huge tree which demolished his car, and received fatal injuries....Folk died en route to the hospital."

Later in the article, it noted that Folk was the first auto racing death in several years, so the heading within the article, "First Auto Tragedy," is somewhat misleading. The article continued, saying that Bo had "disregarded the caution signals from both the starter and the assistant, who tried frantically to slow him down on his first run down the brick straightaway."

I couldn't help but feel that this article was a reminder that racing is and had always been a very risky business, and maybe that is part of its appeal—pushing the envelope and cheating death...most of the time. Still, the article was complimentary to the promoters of the race itself. The event was well attended, and it was very well run, "being the most creditable of any [race] ever handled in the Jacksonville sector." Jacksonville had been put back on "the speed map and along with Tampa and DeLand, this city can take 1934 laurels for auto racing in the state."

AS TIME MARCHED ON, several oval tracks became popular on the Jacksonville racing scene. There had been oval tracks before around Jacksonville. The Moncrief Park Racetrack was where Claude Nolan beat the aeroplane in 1910, but the track was primarily for racing horses. It ran its last race on April 1911. The fairgrounds at Camp Foster lasted only until about 1940, when the naval air station came into existence, but in 1947, Speedway Park, sometimes called Jacksonville Speedway, was opened on Jacksonville's Westside.

This place was owned by Pat Patrick, who also owned Pat's Drive Inn on Normandy Boulevard. The track was a half-mile dirt oval track at Lenox Avenue and Plymouth Street, and if all the articles I have read about it are true, it was "rickety" and had a "seedy" sort of reputation. It was a going concern until 1973, when it closed. The property was then sold, and after much difficulty preparing the contaminated site, it became the grounds of a housing complex in 2003 called Lenox Courts Apartments.

In 1951, NASCAR began to include Speedway Park in its Grand National Series schedule occasionally. NASCAR races were held there in the 1951, 1952,

1954, 1955, 1961 and the 1963 seasons. According to an article posted on the *Metro Jacksonville* website and written by Kristen Pickrell, on December 1, 1963, the last NASCAR-sanctioned race would be held there.

After this particular race, the outcome would be shrouded in controversy that persists to this day, as Pickrell described:

> [Richard] *Petty had been leading for over 100 laps. Everyone thought he had the win, easy, until his steering broke, likely due to the conditions of the track.* [Wendell] *Scott took the lead with 25 laps remaining as Petty fell behind. He crossed the finished line in the lead after lap 200, but there was nobody waving the checkered flag. Two laps later, the flag was waved—for Buck Baker.*
>
> *After complaints, the race steward went back to realize two of Scott's laps were missing. Baker's title was taken, and Scott was given the title of winner, followed by Baker in second, and Jack Smith in third. Because Scott's win was delayed, he never received the trophy. While he may have never been a trophy holder, Scott had won, making him the first African American man to win a NASCAR event.*

An article ran on June 27, 2010, in the *Florida Times-Union*, written by Don Coble and titled "1963 NASCAR Controversy: Racing or Race?" In it, Coble said that after 495 races and a thirteen-year NASCAR career, Wendell Scott had twenty top-five finishes. "His breakthrough race, and the only NASCAR event he ever lead, was in Jacksonville."

Sadly, the events surrounding Scott's breakthrough NASCAR race in Jacksonville looked suspiciously like he had been denied the trophy because of his race. In 1963, that possibility was entirely plausible, but according to most accounts, Scott remained gracious about it all. Coble said that Scott was eventually inducted into the International Motorsports Hall of Fame at Talladega in 1999 and that Scott's helmet, racing suit and a door from his race car are on display at the NASCAR Hall of Fame in Charlotte, North Carolina. Scott spent the rest of his life working as a mechanic. He died in 1990 after a long battle with spinal cancer.

Irwin "Speedy" Spiers (now ninety-one and whose real first name is Irwin, but no one calls him that) was another dynamic force in the local racing industry, and he was even present when the Scott incident occurred. Spiers told me in a telephone interview that he had a list of sixty-five drivers who had raced for him in his cars, like Fireball Roberts and Leroy Yarborough. He himself had raced more than one thousand races, he told me. He did not care for drag racing:

> *It didn't last long enough for me—you spend so much time preparing, and then it's over. Besides, I could make more money on the dirt tracks like the old Jacksonville Speedway. But I raced all over—all along the East Coast to the Mississippi River. Sometimes you made a good living at it, and sometimes you didn't. Still, I loved it, and I even budgeted speeding ticket money into my racing expenses since I was likely to get two on the way to the race and two coming home. It got so that the policemen would say, "See you next time!"*

What I found even more interesting than Speedy's racing career was his recollections of life in Jacksonville in those midcentury automotive days. "Kids needed something to do, so they would drag race up and down Main Street in the cars they had souped up, and the police would look the other way. So long as we didn't hurt anyone, it was okay, but if there was ever an accident, they would really crack down."

He also told me that after the war, when the war effort was winding down, there was an influx of young veterans needing work. They could work for the shipyards for $60 or $70 a week, or they could run moonshine for $300 a week. He chose the latter for a while. "North Florida was the moonshine capital of Florida," he said. "Macclenny was the hub of the operations. And the way the revenuers would catch moonshiners was to look for wild hogs that were stumbling about drunk in the thick swamps from eating the sour mash they would get into."

Speedy made a good living running moonshine, and he was only caught once. "I was really lucky. I just received a slap on the wrist, but I would not have been so lucky a second time. I'd have spent a year in Marianna."

Of course, talking with Speedy reminded me that the Jacksonville Speedway was quite popular when I was growing up here in Jacksonville. WAPE, the teenager radio station of choice, was always advertising the races on Thursday and Friday nights. There were several races a night, and many of my contemporaries would go there when they could. And according to Kristen Pickrell's article, Lynyrd Skynyrd band members were often seen there.

Another oval track opened in 1968. The Gator Parkway (New Jacksonville Speedway) opened at 186 Pecan Park Road. Not only did it have a half-mile clay oval, but it also had an eighth-mile drag strip. In 1973, the half-mile track was replaced by a smaller, paved oval. In 2004, the place closed entirely.

As I was researching and starting my first drafts of this book, I was able to interview Bobby Tyer, a local driver who had frequently raced at Speedway Park/Jacksonville Speedway. And as often happens as I grow older, it wasn't much more than month later that I was in attendance at his funeral.

"He smelled of metal," Shelby Nolan said of the first man she had ever loved: her foster father, Bobby Tyer. "He used to tell me not to hug him when he first came home from work, at least not until he had showered, but I didn't care. I hugged him anyway because I loved that smell—that smell of hard work and metal."

Her moving tribute to Bobby Tyer brought tears to the eyes of all those assembled to pay homage to an extraordinary man. Even the stoic men who had raced cars with him, flown planes with him or built engines with him had red-rimmed eyes, and the place was packed with many people standing in the back. Everyone who had known Bobby was there to say goodbye to a bigger-than-life person who could race cars, build and fly airplanes and then come home to fragile babies in need of his wife's medical care and his loving affection.

I am so grateful that I was able to interview Bobby Tyer before he died. We had gotten together on a rainy afternoon at his home in Mandarin, and as we sat at his kitchen table, he took me on a roundabout narrative journey of racing in Jacksonville, a passion of Bobby's, particularly in his youth. He even made it possible for me to interview Speedy Spiers.

Jan, his wife, was there, and she began by announcing that Bobby was not a scholar. "When Bobby was in first and second grade, he'd just crawl out of the window and walk home. He did not like school." What Bobby did like was "messing with cars and then racing them." Thankfully, he eventually attended Technical High School, where he could put his talents to good use.

Suddenly, Jan left the room and returned with an armload of blown-up photographs of Bobby looking like a character out of the movie *Grease* standing next to some of his best cars: 1/2 Pint, a 1933 Chevrolet and Sugar Bugger/P5, a 1937 Ford. You could see why Jan would have fallen for such a powerful, handsome character, although these pictures were from his Speedway days before they met at a track in Brunswick, Georgia.

Bobby loved to race, and Jacksonville was a great place to do it. He said that the weather was good here and that there was a dirt track off of Lennox Avenue that held races every Saturday and Sunday. That had to have been the old Jacksonville Speedway he was talking about. But he didn't confine his racing just to Jacksonville. "I often had to hock a tire to have enough money

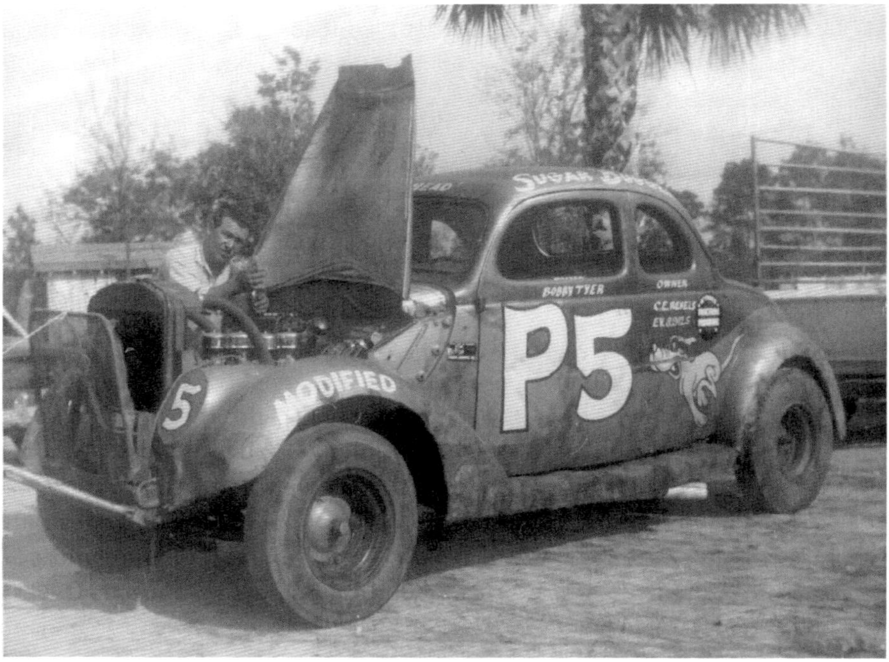

Bobby Tyer with Sugar Bugger. *Courtesy of Jan Tyer.*

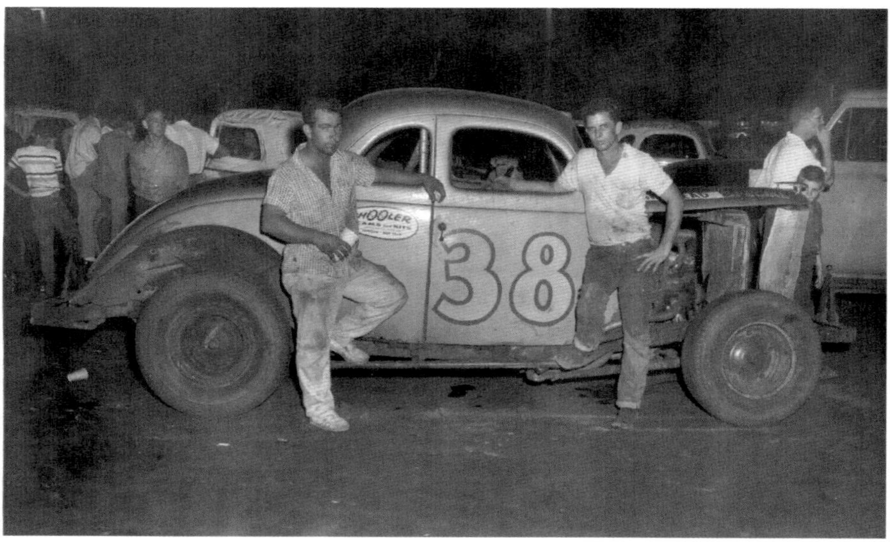

Bobby Tyer (*left*) between races. *Courtesy of Jan Tyer.*

to drive up to Valdosta to race," he told me. "Then, with my winnings, I'd buy the tire back when I got home."

At Bobby's funeral some weeks later, many people came forward to speak what they knew of Bobby. Words and phrases like *master builder*, *inventor*, *life of the party*, *sky diver* and *speed freak* were used to describe this man who made working with his hands an art form. For certain, his mechanical genius was something to behold. It occurred to me that "Tyer" was indeed a most appropriate name for this man to have inherited. Perhaps it was no cosmic accident that a man who loved racing as much as Bobby had would have a name to go with that vocation.

There is another powerful story to tell about racing in Jacksonville, and it concerns a pair of racers who put Brumos and Jacksonville on the racing radar. Peter Gregg and Hurley Haywood raced together for Brumos in the '70s and '80s. Not only is theirs an important story to the history of racing in this area, but it is also the stuff of Hollywood movies.

Peter Gregg had it all: good looks, a Harvard education, a wife and two sons, race cars and the talent to win many racing championships. In an article titled "Spirit that Lives," written by Don Coble for the *Florida Times-Union*, Gregg was described as a native New Yorker who was stationed in Jacksonville with the navy and never left:

> [Gregg] *bought Brumos* [after the death of Herbert Brundage], *which included three dealerships, in 1968 and won 47 races on the International Motor Sports Association* [IMSA] *circuit between 1971 and 1980. He also won the 24 Hours of Daytona three times.*
>
> *Gregg offered Haywood a spot on his team after Haywood beat him at a local autocross race, and Brumos drivers Gregg, Haywood and Brian Redman combined for eight IMSA championships and four Sports Car Club of America Trans-Am championships.*

If the articles I have read on Peter Gregg are any indication of the reality of his personality, he was first and foremost a perfectionist. He was mentioned a number of times as "Peter Perfect." In his racing life, this perfectionism served him very well, and he became very well known throughout the city, the nation and the world. In his personal life, this perfectionism contributed to or was the result of a manic depression for which he was being treated. In an article titled "Conversations with Peter Gregg," written by Mark

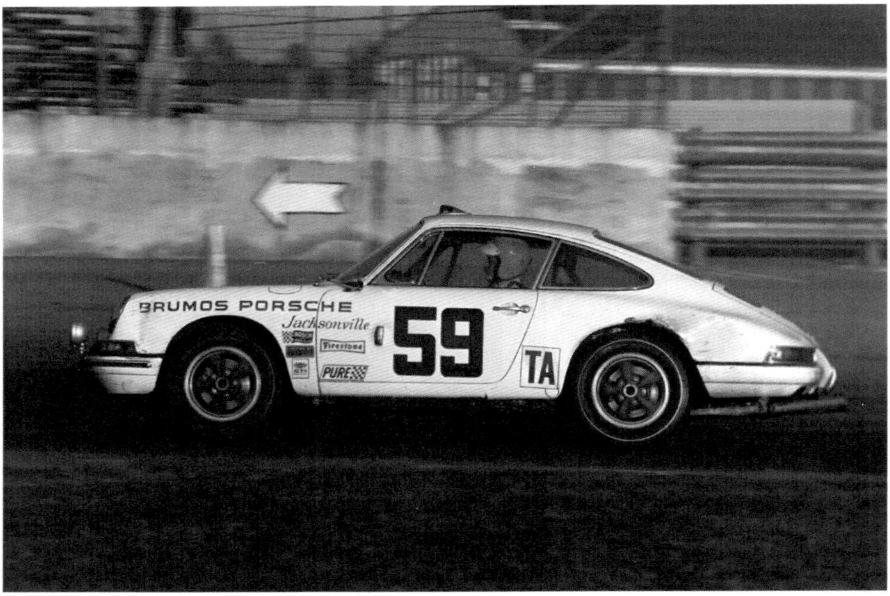

Peter Gregg in Brumos 911 Porsche Daytona, 1969. *Courtesy of Bill Warner.*

Raffauf and published on the website Porsche Road and Car on April 21, 2017, it noted that in June 1980, not long before the 24 Hours of Le Mans Endurance Race:

> *Peter came into our pit and he was all banged up with bandages on his head, and bruising on his face and arms. He was walking slowly, and those of us who knew him fairly well, came right over and asked what had happened. He told us he was out of the race, he had been on his way to the circuit in a road car when a French farmer had pulled out in front of him with a tractor, forcing him off the road into a ditch. He was suffering from dizziness and blurred vision, was not going to drive and was, in fact going home early. Since he knew us from IMSA he wanted to come by and wish us well before he left. Very classy, we thought at the time. We wished him well, and were glad he was not more seriously hurt.*

The reality was that Gregg's injuries were far more serious than he was letting on. He had a concussion with double vision, and he was not cleared by the officials to race. In December 1980, for reasons we can only guess, Gregg committed suicide on a Ponte Vedra Beach and shocked the world.

He left behind his new bride of nine days, Deborah Jane Marrs, to continue running the Brumos dealership and to wonder. The article went on to say:

> *Deborah Gregg inherited the dealerships, located on Atlantic Boulevard and dealing in exotic cars like Porsche, Audi and Mercedes-Benz, but eventually she sold it to* [Bob] *Snodgrass and Dan Davis. She dabbled in club racing but essentially has stepped away from the lifestyle and the responsibilities her husband left behind.*

Still, this tragedy could not diminish the outstanding racing accomplishments of Gregg's life, and there were many wonderful racing moments for Gregg and his driving partner, Hurley Haywood. The article quoted Haywood as saying, "What motivates us, even after Peter died more than 20 years ago, is his passion. Peter Gregg was passionate about his racing, and that rubbed off on everyone. He demanded that same kind of passion, and it's something that still drives this race team."

Hurley Haywood was of a very different temperament than Gregg. In an article written by Simon Taylor in the July 2016 issue of *MotorSport* titled "Lunch with Hurley Haywood," Haywood revealed himself to be "quiet, self-contained."

As a kid, Haywood used to vacation on his grandmother's farm just west of Chicago, and he learned to drive there. In 1967, he was in college in Jacksonville and had a Corvette that he used to race in autocrosses. In one race, he beat the owner of Brumos in Jacksonville, Peter Gregg, who said to him, "You must be pretty good to beat me." Gregg invited Haywood home

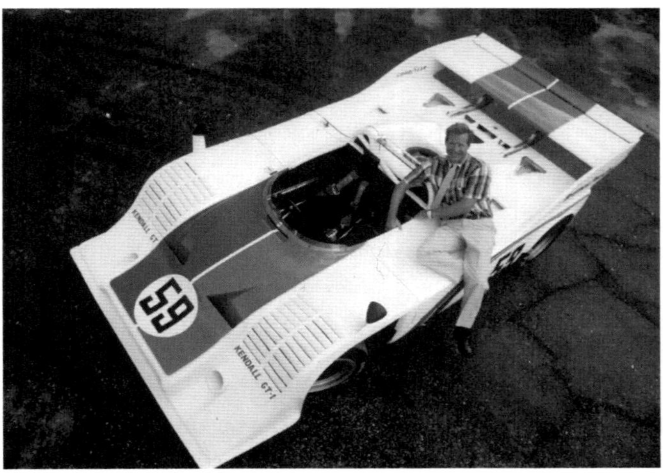

Hurley Haywood in 917-10 Porsche Can Am Car. *Courtesy of Bill Warner.*

to dinner, and that was how they met and formed a racing relationship. Haywood noted:

> *After that he took me to the Savannah Raceway to give me my first run in a proper race car, a 911S. By the end of the day, I was faster than him, and he said, "Looks like I'm not going to teach you anything about driving, but what I can teach you is the mind-set of a professional racing driver. And you and I are going to do the Watkins Glen Six Hours next month."*

With Gregg's talent and mentoring, they were able to win first in their class at the Watkins Glen Six Hour Race, and they finished eighth overall. One month later, Haywood was drafted and on his way to Vietnam, stationed about one hundred miles south of Hanoi. It was in the service that he also learned important lessons that would serve him well in racing. He said that he learned "the protocol of when and where to do what and how to be very efficient with my time and my energies."

When Hurley was discharged, he went to work with Gregg and the Brumos team, and after a baptism by fire of learning the ropes, Haywood realized that Gregg left nothing to chance. From the precise car colors of red and blue stripes on a field of white to the spotless uniforms of all the crewmembers (even mechanics) to an atmosphere short on compliments and an expectation of excellence from everyone, "Brumos had to be in a different league."

And it was this league-of-their-own mentality that helped Haywood win the 24 Hours of Le Mans Endurance Race three times in three different decades, the first one being his first Le Mans race ever. He would win the 24 Hour Daytona Endurance Race a record five times. He would win the 12 Hour Sebring Endurance Race two times, and he would rack up forty wins in major sports car races during his racing career, which spanned forty-three years.

After Gregg's death, life and racing did go on. Haywood would continue racing until 2013, and he would become VP of Brumos. Then, in 2016, Brumos's two Mercedes dealerships, two Audi dealerships and one Porsche dealership were purchased by the Field Auto Group out of Chicago. The name Brumos, however, was not sold and is still owned by Dan Davis.

At the present time, a new Brumos facility is being built into which the Brumos Car Collection will be moved. The facility plans to showcase this collection of race cars, have a diner and shops and will have closed tracks nearby for them to "exercise the cars" without taking them on public roads.

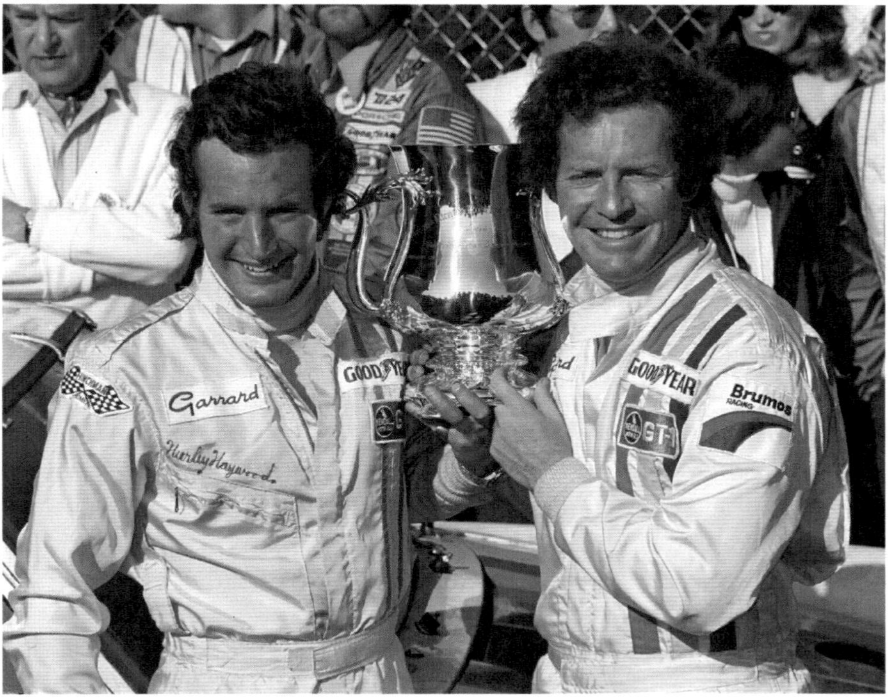

Hurley Haywood (*left*) and Peter Gregg winning Daytona, 1969. *Courtesy of Bill Warner.*

They were even able to get the City of Jacksonville to allow the address to be 59 Lee Street in honor of the numbers that were on the race cars of the Brumos team. A book about Heywood's life is scheduled to be released in 2018, and there is a documentary by Derik Dodge that is soon to be released about the careers of Gregg and Haywood.

It would seem that these are fitting memorials to the "perfectionism" of a man called Peter.

I suppose that when all is said and done, racing automobiles is much more than just proving who is fastest. Maybe it is more about what little girls remember of their fathers—the smell of hard work and metal. At the very least, racing has provided countless hours of blood, sweat and enjoyment, for racers and spectators alike, and Jacksonville has provided ample opportunities for this enjoyment to exist. Whether it was Claude Nolan flying faster than an aero plane at Moncrief Park on the Northside, Bobby Tyer speeding around the track at Speedway Park on the Westside, Bob Burman flying down Atlantic Beach to Pablo Beach and back or Peter Gregg and Hurley Haywood representing Jacksonville in worldwide racing arenas, all of these racing enterprises have been grand fun. Even though other places like Daytona have superseded Jacksonville as a place where races can be conducted, Jacksonville can be certain that it and some of its favorite sons have a place in the annals of racing itself.

LOVING OUR AUTOMOBILES

CARS & COFFEE, MAIN STREET CRUISE AND CONCOURS D'ELEGANCE

I first came to Jacksonville in 1951 as a naval photographer assigned to VC-62, a Photo Aviation Squadron at NAAS Cecil Field. Part of the squadron's history was this car, a 1941 Packard Convertible, as it was only sold and resold to members of the squadron. Then, as each owner left the squadron or went on detached duty, they would sell the car to someone else at the same price that they paid for it, ranging between $200 and $300—provided it was in the same condition it was when they purchased it. I was fourth in line to own it (for $300), as the squadron was newly commissioned in 1949. When I received orders to go on detached duty aboard the aircraft carrier Franklin Roosevelt, *I in turn sold it to another member of the squadron for the same amount that I paid for it. The provisions of sale were that when the registration was transferred to the new owner, the bill of sale and a copy of the registration would be kept in the Executive Officer's Office safe until all the monthly payments were deposited to my bank account. The buyer would show the Executive Officer the proof of monthly payments, and when the final payment was made, the buyer would take possession of the bill of sale.*

I left the navy and returned to New Jersey, where I stayed for two years, and then came back to Jacksonville to start my career in television at WFGA-TV Ch-12. During my time in Jacksonville I once saw the VC-62 Packard parked downtown. Whoever owned it had customized it by removing the convertible cloth roof and replacing it with a black, sleek fiberglass roof; however, I still recognized the rest of her.

—Herb Gold

It has never ceased to amaze me how much our cars mean to us. Aside from the utilitarian aspects of cars, there is the pleasure that comes from fabulous adventures. But there is also something about getting your hands dirty to make an engine run properly or making a rusted-out old bucket into a thing of beauty. Perhaps this affection has to do with distant and happy times in our automotive lives that we are constantly trying to re-create. Maybe it is just something to do.

On January 14, 2017, a Cars & Coffee event was held at the Avenues Mall located at US 1 and Southside Boulevard. It was the first such event held at this location. One month earlier, the organizers held the event at the Florida Times-Union Building parking lot on Riverside Avenue.

My husband, Hardy, and I headed for the Avenues that January morning in drizzly, cold weather. When we rounded the corner to get to the entrance of the mall parking lot, there were cars and owners as far as the eye could see. The entire east parking lot of the mall was filled with cars and their proud owners despite the wet weather. Excited young boys scurried from car to car, taking pictures with their cellphones. Older men wandered the wet parking lot admiring the renovated marvels from the past—Packards and Plymouths and Chevys and Fords. Then there were the hot rods and the latest shiny offerings of Ferrari and Lamborghini. Everyone was friendly and talkative and seemed to be happy to be there, even in the drizzle that didn't let up until ten o'clock or so.

It was wall-to-wall at another Cars & Coffee event for car enthusiasts. It took place at EverBank Field on October 14, 2017. Mustangs and hot rods, Rolls Royces and VWs—it was an amazing display of wonderful cars and cordiality. Complete strangers would carry on long conversations about design or travels. By the time we had seen it all, we felt a warm glow that was more than just the unseasonable heat. The entire event was pleasant. Whether it was the smallest car-loving kid with his quick "I like your car" or the most-seasoned driver who could offer advice about a problem, this event was a grand time. I would call it an unqualified success.

Jacksonville provides quite a number of automotive "sharing" experiences. Along with the Cars & Coffee at the Avenues and Everbank Field, there is the Main Street Parade of Vintage Cars, which runs every third Saturday evening of the month. Here, cars of all makes and models drive up and down Jacksonville's Main Street in an effort to re-create the *American Graffiti* cruising experience. With its terminus being the Krystal Restaurant—which is not really at the original location anymore—the route allows drivers the opportunity to parade their cars in front of an accepting audience that lines

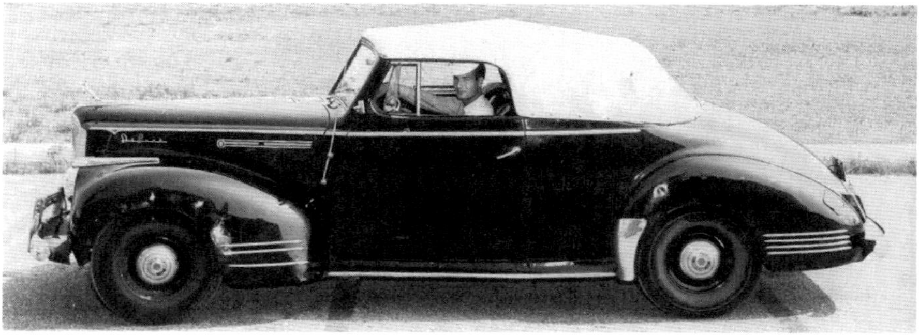

Above: 1941 Packard Roof Up at NAAS Cecil Field, 1952. *Courtesy of Herb Gold.*

Below: Future car enthusiast, Cars & Coffee at Avenues Mall. *Author's collection.*

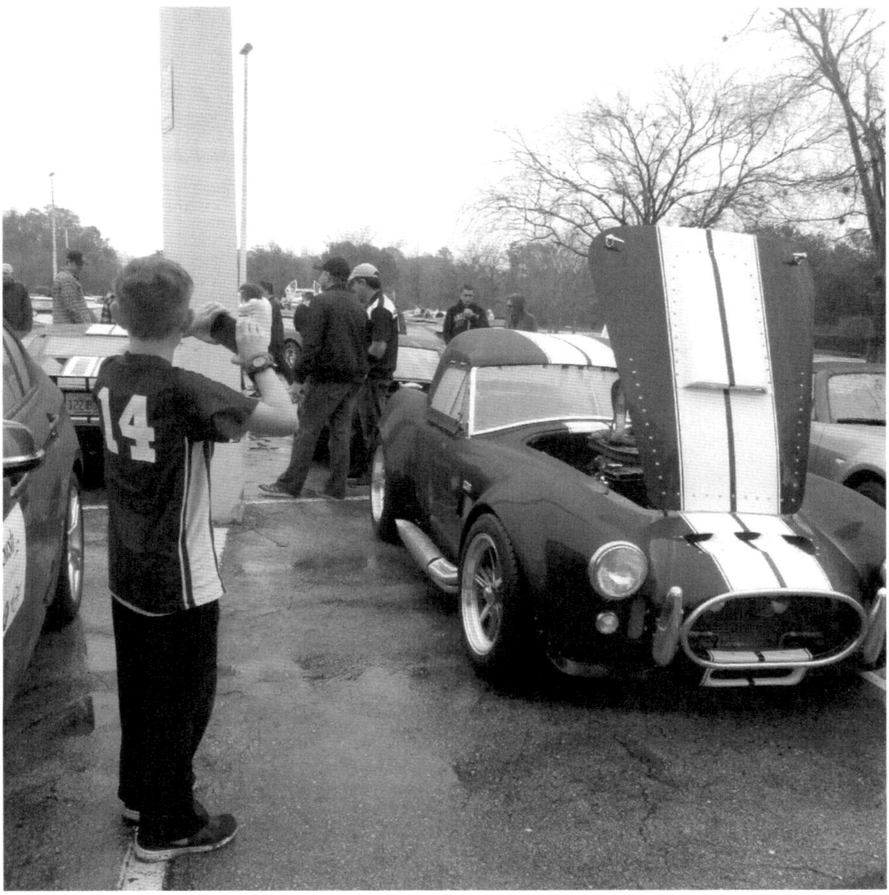

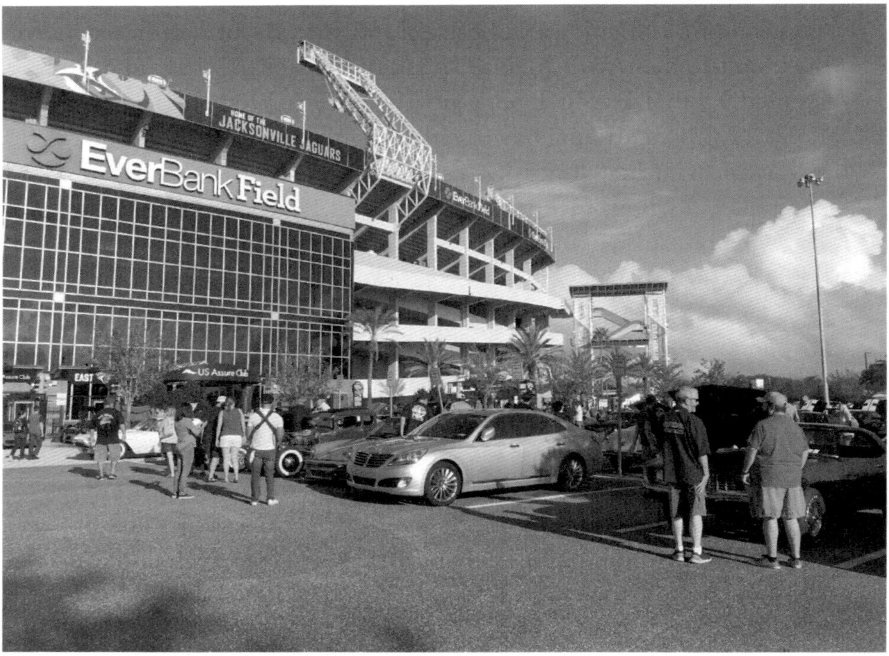

Everbank Field's Cars & Coffee. *Author's collection.*

the street. An AM radio station plays oldies, taking everyone back to the days of our youth. Food trucks bring wonders to eat and smells to savor.

What is most interesting to me is how many people participate in the Main Street Cruise. While many drive their rides up and down the thoroughfare, slowly and sometimes loudly, others position their cars along the "main drag" for those who like to walk past and take in all the bling and shine of the pampered rides of participants.

Of course, the biggest of all the "show off venues" is the Amelia Island Concours d'Elegance. It is French for, roughly, "Parade of Elegance," and when the beautiful cars roll by, it is indeed a lovely sight! It began in 1996, and it takes place every spring at the Ritz-Carlton in Amelia Island.

This internationally known, award-winning event has truly put the North Florida on the car lovers' map. It was the brainchild of an incredible force of nature in the world of automobiles: Bill Warner, someone I've mentioned several times so far. Warner, who is the founder of the Concours d'Elegance, told me that he grew up pestering all the dealerships on Philips Highway near his home. Before he could even drive, he'd bicycle each week to their showrooms to get the latest brochures and check out the latest models.

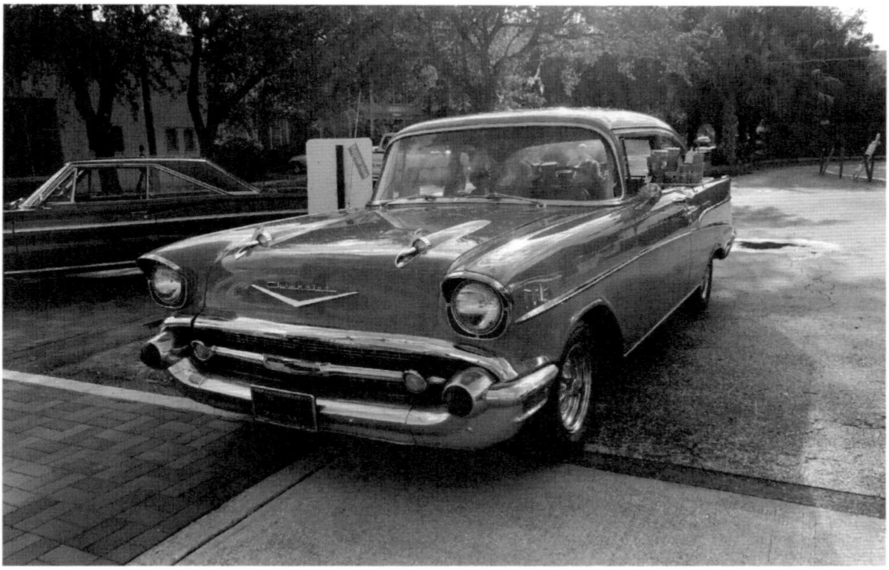

Main Street Cruise; '57 Chevy with food tray on the window. *Author's collection.*

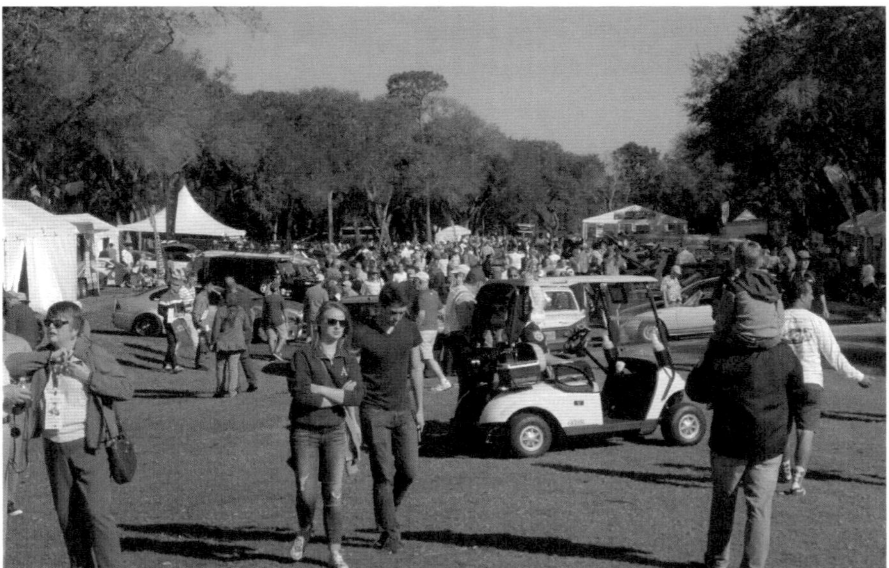

Concours d'Elegance, 2017. *Author's collection.*

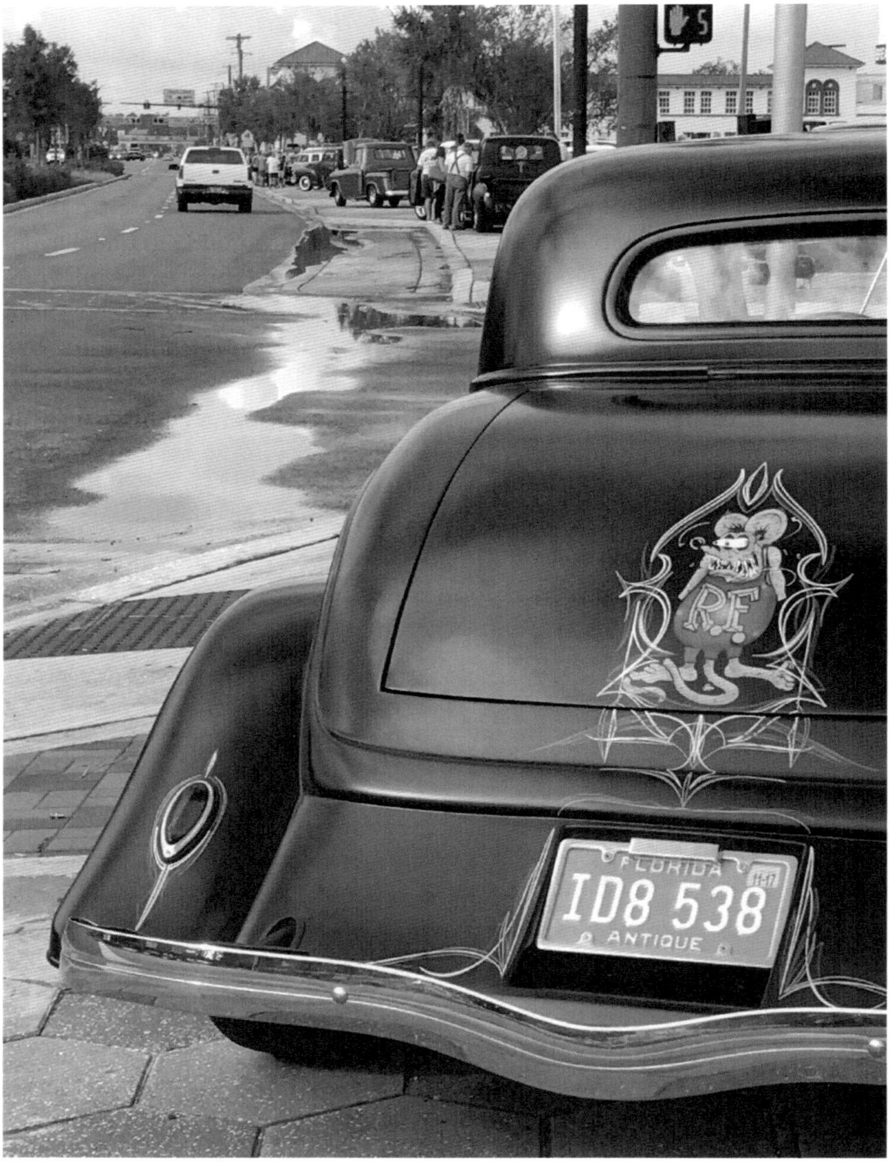

View down Main Street. *Author's collection.*

Warner has always loved cars, but he also loves going fast and has quite the résumé to prove it. According to the website of the Amelia Island 2019 Concours D'Elegance, Warner has won "the IMSA Media Challenge in 1985, 1986 and 1987, driving a Chevrolet Camaro in the Firestone Firehawk series with Tom Nehl and Buddy Norton. He also co-drove with Norton, Sean Hendricks and Larry Hendricks to win the SCCA 24 hours of Mid Ohio Escort Endurance race. It was the first professional win in history for an automatic transmission production car. He also is a veteran of two Cannonballs. For the past two decades, he has raced the ex-Group 44/Paul Newman Triumph TR-6 in vintage events [in which he allowed me to sit] and has won both the 1997 Skip Barber Cup and the 1999 John Kelley Memorial Cup."

One accomplishment in which Warner takes great pride is being part of two of the first Cannonball Runs of the '70s. Earlier in the previous century, the Cannonball Runs were sea-to-sea endurance runs named for Erwin "Cannonball" Baker, who first rode his Indian motorcycle from Los Angeles to New York City in 1922 in six days, twenty-two hours and fifty-two minutes.

Warner's first endurance race took place in a 1971 Porsche, and in 1975, many years after Baker's run. Warner and seventeen other people met at the Red Ball Garage on Thirty-First Street in New York. The one who got to the Portofino Inn in Redondo, California, first would be the winner. It must have been exhausting but amazingly exciting!

Along with his work as the founder and co-chairman of the annual Concours d'Elegance in Amelia Island, Warner is also president of H.C. Warner Inc., an industrialization filtration company. When he isn't doing that, he's co-writing books like *Cuba's Car Culture* with Tom Cotter, which naturally required a trip to Cuba. And then there was the Amelia Island Concours d'Elegance winning the International Historic Motoring (IHMA) Event of the Year Award in 2013 and 2016. This award is one of the most prestigious awards for such venues.

When my husband and I went to the Cars & Coffee segment of the Concours d'Elegance in Amelia Island in 2017, I was blown away by the enormity and the classiness of the event. There were so many interesting people and fabulous automobiles, and everything moved flawlessly together with grace and precision.

Since a very bad weather front was threatening the area and was due to arrive on Sunday, the powers-that-be canceled the Sunday activities and moved all of them to the Saturday, March 11, 2017. Both the Concours d'Elegance and the

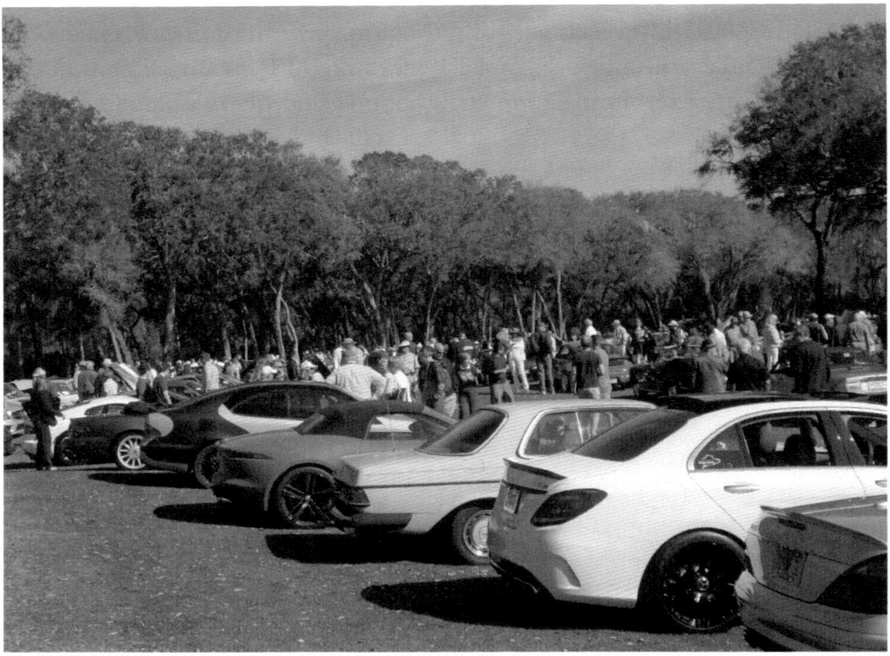

Concours d'Elegance, 2017. *Author's collection.*

Cars & Coffee were held on the same day on different parts of the complex of the Ritz-Carlton Hotel on Amelia Island. Hardy and I were amazed at how seamlessly the change in schedule was executed. We took a shuttle from the airport to the Concours area and enjoyed a chamber of commerce day as we promenaded across wide expanses of well-manicured golf courses to enjoy looking at the wide variety of cars at the Cars & Coffee.

Very few old cars were among the displayed automobiles. Most of the cars were very expensive—one Bugatti Veyron was $1.6 million, and another was available for $3 million. There were also Mustangs, Corvettes, Bentleys and Ferraris lined up side by side—in three lines sometime—over at least two holes of the golf course. It was magnificent!

The Concours always causes quite a stir with its impressive offering of cars to enjoy. In 2016, Dan Scanlan of the *Florida Times-Union* wrote about some incredibly remarkable cars featured that year. One of the most popular was Janis Joplin's 1964 Porsche 356 C 1600 Cabriolet. The 1964 model was psychedelically painted and caused quite a ruckus. *Chasing Classic Cars* host Wayne Carini's was there that year as well with his 1949 Buick Roadmaster convertible, one of two used in the film *Rain Man*, starring Dustin Hoffman and Tom Cruise. It, too, was quite popular.

What one really needs to know about the Concours is that it is much more than just an opportunity to show off cars. It raises hundreds of thousands of dollars for charities, most especially Community Hospice of Northeast Florida, the Navy-Marine Corps Relief Society of Mayport, Spina Bifida of Jacksonville and the Shop with Cops program in Fernandina Beach.

As my husband and I began our journey through the maze of cars, we could hear "The Star-Spangled Banner" filling the air as everyone stopped and almost reverently sang along. Then, five vintage aircraft flew overhead. There was a fashion show that was part of the Concours Parade of Automobiles, and it was hard to believe that bad weather was fast approaching. It was an all-around wonderful day, topped off by lunch at Barbara Jeans at the less traveled part of Amelia Island. The rain held off until we made it all the way home.

The final event that I wanted to mention was the inaugural Cars & Coffee event at the HSR (Historic Sportscar Racing) Classic 24 at

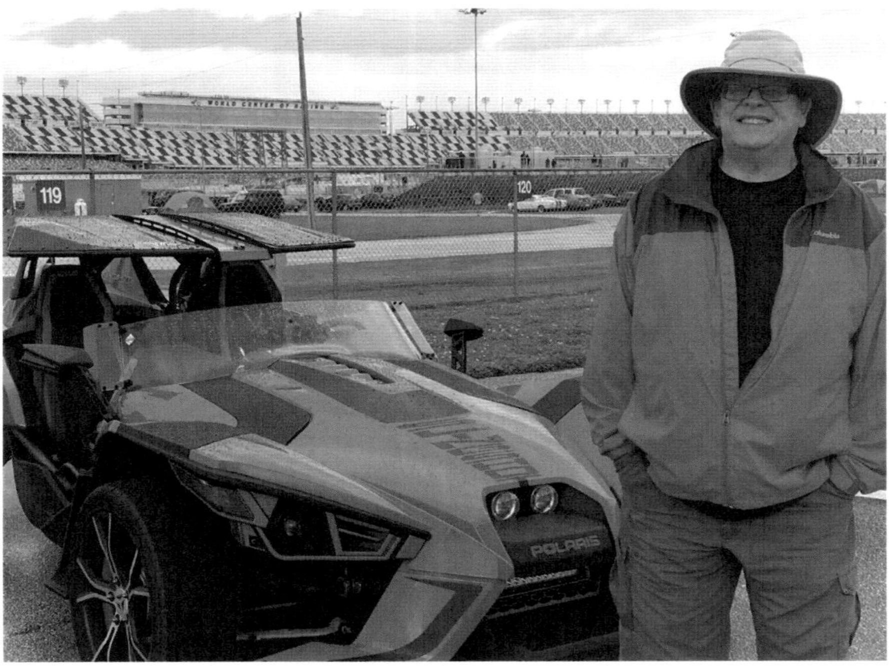

Hardy Fletcher at Daytona International Raceway for Cars & Coffee, 2017. *Author's collection.*

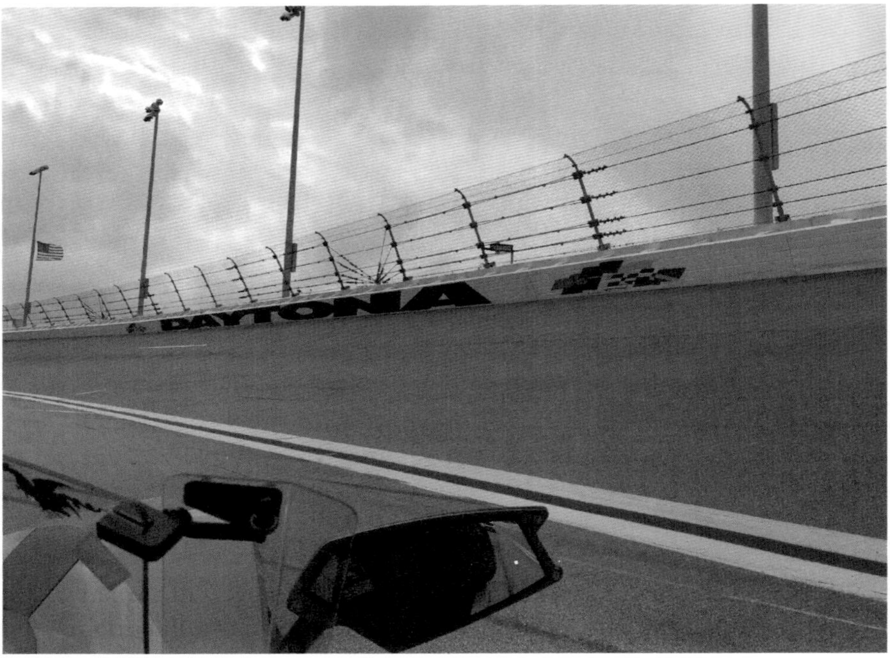

Parade lap at Daytona International Raceway. *Author's collection.*

Daytona International Raceway on November 11, 2017. This wonderful event was a privilege and one of the highlights of my writing this book. My husband and I had been invited to participate in a parade lap by being part of the Cars & Coffee at the Classic 24. We had ridden through the rain to Daytona that morning in my husband's Slingshot. The trip was uncomfortable and difficult since his Slingshot is a three-wheeled motorcycle that functions a lot like a car but is still open to the elements. We were a bit damp when we arrived at the speedway, but our enthusiasm wasn't.

At 11:45 a.m., the parade lap began after a mandatory meeting, and wristbands were given out. It was absolutely thrilling to be able to drive on the very road that countless NASCAR drivers had driven—where Peter Gregg and Harley Haywood had won many races. It was awesome to know that Dale Earnhardt had won here and, sadly, had died here. Sig Haugdhal's dream of a raceway had come to be at this magnificent place! It was all I could do to keep from tearing up and missing the spectacular views.

Bill Warner was not far away, preparing for his race in his 1965 Shelby GT350 in the pits, and at 1:00 p.m., the Classic 24 began. I was incredibly happy, and I couldn't help but wonder what my insurance adjustor father would think of my being part of all this.

CONCLUSION

No matter how you look at it, most of us relate to our cars very emotionally. Cars are so much a part of our existence—from being brought home from the hospital as "infants" to the sad journey to the cemetery. But in between those two rides, automobiles have captured our hearts and imaginations. And considering that it has been fewer than 120 years since Einig's "chug chug" machine puttered about Jacksonville streets, mystifying the populace and terrorizing the horses, we can all be amazed that cars have so rapidly and powerfully affected all of our lives.

I am left with one sterling moment of what makes car events so magical. In all of my journeys to Cars & Coffees and cruises and Concours, I was surprised at the visceral feeling a 1957 Ford Thunderbird created in me as I was wandering the rows of old beauties at one of the numerous events I had attended at the Avenues. I was passing the Forever 21 store (how appropriate!), and all of a sudden there it was! It had the colors of a Dreamsicle, and they really popped in the beaming sunshine. The creamy white of the pristine leather interior blended perfectly with the soft orange of the exterior. It took my breath away. I could almost taste it!

I closed my eyes and could easily imagine myself as a young woman sitting in the driver's seat, ready for a day at Jax Beach. The seats would be filled with me and all my happy friends, the smell of Coppertone wafting through the warm air and mixing with tunes from the Big Ape, WAPE—maybe some "Jailhouse Rock" by Elvis or "Long Tall Sally" by Little Richard. We would

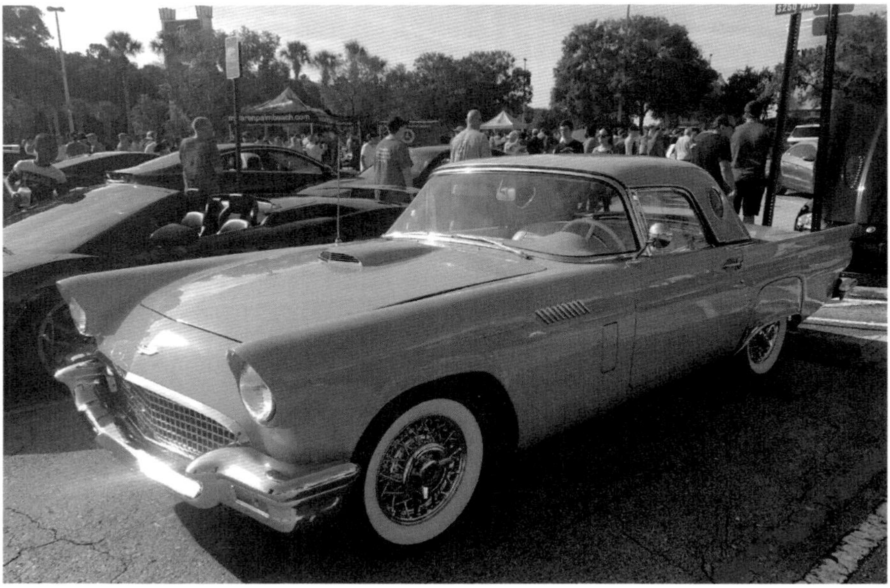

My fantasy Dreamsicle'57 Thunderbird. *Author's collection.*

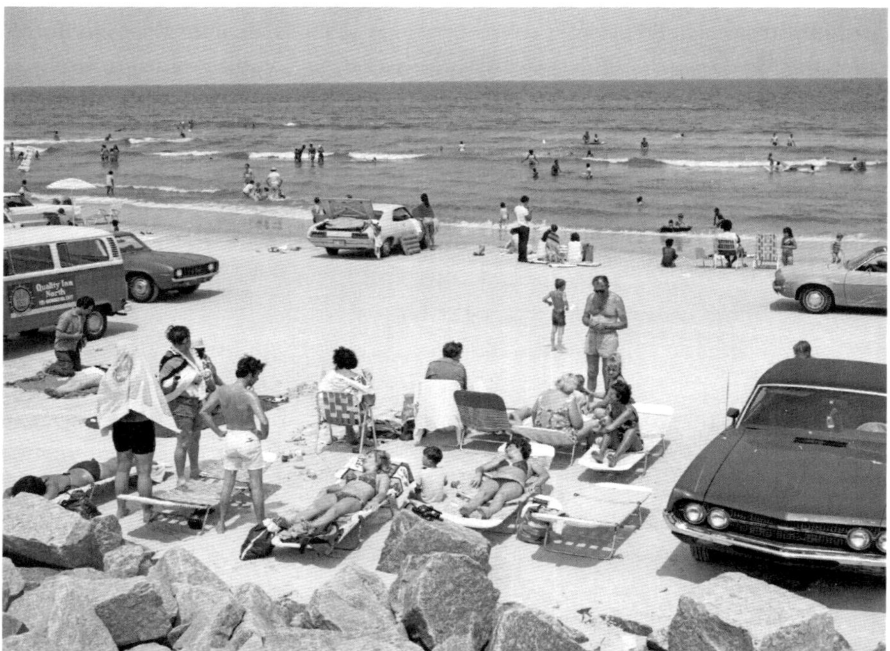

The Jacksonville Beach of my youth. *Courtesy of Florida State Photographic Archives.*

all be singing along and laughing at our amateurish performances as the wind whipped in through the windows.

As I imagined us headed down the highway in search of the perfect sunbathing spot, I realized that most of us in America have our golden car moments. Maybe it is that used car someone's father brought home to the family that seemed "brand new." Maybe it was the first car someone ever bought with his or her own money. Maybe it was the car in which someone picked up a date to go to Pennyburger's and then to a drive-in movie. Maybe it was the car that someone rebuilt to its former glory or the hotrod someone souped up to go racing. No matter what, we Americans have made cars an integral part of our culture, and Jacksonville has been part of it all from that day when the *Florida Times-Union & Citizen* announced that a car was coming to town!

BIBLIOGRAPHY

Amelia Island 2019, Concours D'Elegance. https://www.ameliaconcours.org/Bill-Warner.aspx

Automotive News. "IN THE BEGINNING." Timeline (September 26, 2006): 6–36.

Bachmann, Andrew, and Maria E. Mediavilla, in association with the Jacksonville Historical Society. *Jacksonville Revisited.* Charleston, SC: Arcadia Publishing, 2007.

Breittenstein, Bill. "Wild Sam Purvis Takes 100-Mile Automobile Race at Camp Foster." *Florida Times-Union*, April 2, 1934.

Brundage, Jan. "Racing History." E-mail to author, November 6, 2017.

Carney, Phil. "The Brumos Story." *Excellence: The Magazine About Porsches* (February 2004).

Celebrating 80 Years of Tradition and Excellence. Jacksonville, FL: Scott-McRae Group, 1996.

City of Jacksonville. "About Jacksonville: A Rich & Vibrant History." www.coj.net/about-jacksonville/history.

Claude Nolan Cadillac: Celebrating 100 Years. Jacksonville, FL: Claude Nolan Cadillac, 2005.

Coble, Don. "1963 NASCAR Controversy: Racing or Race?" *Florida Times-Union*, June 27, 2010.

———. "Spirit that Lives." *Florida Times-Union*, January 25, 2004.

Crane, Keith E., publ. and ed. *Automotive News* (September 23, 2006): 3.

Daily Record. "Fifty Years Ago This Week." March 26, 1962.

Davis, Ennis, and Robert Mann. *Reclaiming Jacksonville: Stories behind the River City's Historic Landmarks.* Charleston, SC: The History Press, 2012.

Davis, T. Fredrick. *History of Jacksonville, Florida and Vicinity, 1513 to 1924*. A facsimile reproduction of the 1925 edition. Gainesville: University of Florida Press, 1964.

Florida Times-Union. "ANOTHER AUTOMOBILE REPORTED MISSING." May 30, 1910.

———. "ATLANTIC BOULEVARD CELEBRATION." July 29, 1910.

———. "AUTOISTS FIGHT BILL." May 20, 1903.

———. "AUTOMOBILE DRIVERS GIVEN WARNING ABOUT BEACH NEAR JETTY." June 22, 1910.

———. "AUTO PARADE TO BE A FEATURE." October 16, 1903.

———. "BUCKMAN BRIDGE IS OPEn." May 1, 1971.

———. "Claude Nolan Dies After Brief Illness." February 3, 1943.

———. "DEATHS: C.A. Brown, Claude Nolan President." November 22, 1976.

———. "First Cars Cross Isaiah Hart Bridge." November 3, 1967.

———. "Florida's Famous Road to the Sea." July 29, 1910.

———. "FORD OFFICERS MOVE IN PLANT." August 29, 1924.

———. "Gordon Thompson Funeral Scheduled Saturday." April 4, 1975.

———. "MANY AUTOMOBILES IN JACKSONVILLE/MACHINES REGISTERED WITH RECORD NUMBER 166." November 4, 1905.

———. "Mathews' Dream Realized as Great New Bridge Over St. Johns River Dedicated." January 2, 1953.

———. "New Acosta Bridge Markers to Be Erected Here Monday." August 18, 1949.

———. "New Gateway to City." September 23, 1988.

———. "Span Dedication Is Slated Today." June 7, 1954.

———. "Wheel-to-Wheel and Hub-to-Hub." April 1, 1934.

———. "YESTERDAY'S PARADE." November 5, 1903.

Florida Times-Union & Citizen. "BRIEF LOCAL NEWS." August 12, 1900.

———. "BRIEF LOCAL NEWS." February 2, 1901.

———. "THE FIRST HORSELESS CARRIAGE RECEIVED YESTERDAY." April 30, 1896.

———. "FIRST IN FLORIDA/C.A. CLARK IMPORTS A MODERN HORSELESS CARRIAGE." January 5, 1900.

———. "IN A HORSELESS CARRIAGE JOHN EINIG WILL HAVE RIDE ON THE FOURTH." July 2, 1899.

———. "JACKSONVILLE DEVASTATED BY A MOST DESTRUCTIVE CONFLAGRATION." May 4, 1901.

Foley, Bill, and Wayne W. Wood. *The Great Fire of 1901*. Jacksonville, FL: Hartley Press, 2001.

The Ford Assembly Area Revitalization Study, 2006. Provided to the Planning and Development Department of City of Jacksonville by ARCADIS (BHR) Project Team. Asset Property Disposition Inc., RKG Associates Inc.

Ford News. "New Jacksonville Plant Completed; Capacity 150 Cars Per Eight-Hour Day." March 1, 1925.

Gold, Pleasant Daniel. *The History of Duval County, Florida*. St. Augustine, FL: Record Company, 1928.

A History of Creating Value. DVD. Scott-McRae Group, 2004.

Kimes, Beverly Rae. "The Real Winner Was the Beach." *Something About Everything Racing*. http://www.jcs-group.com/racin/racing/beach.html. Originally published in *American Heritage* 38, no. 7 (November 1987).

Leslie J. Allen. "The Franchise System." *Automotive News* (September 26, 2006): 6–14.

Pace, Harold W., and Mark R. Brinker. *Vintage American Road Racing Cars, 1950–1970*. St. Paul, MN: Motorbooks International, 2004.

Patton, Charlie. "Mike Shad Paying Physically for His Years in Vietnam." *Florida Times-Union*, January 8, 2017.

Pickerell, Kristen. "Lost Jacksonville: The Story of Speedway Park." *Metro Jacksonville*, November 4, 2015. https://www.metrojacksonville.com/article/2015-nov-lost-jacksonville-the-story-of-speedway-park/page.

Prior, Richard. "Mike Shad Buys Out Bozard." *St. Augustine Record*, November 10, 2005.

Raffauf, Mark. "Conversations with Peter Gregg." *Porsche Road and Car* (April 21, 2017).

Scanlan, Dan. "A Beach Racer that Sliced through the Sands of Florida Is Restored and on Display." *Florida Times-Union*, June 13, 2009.

———. "A Toast to Amelia." *Florida Times-Union*, March 19, 2016.

Southern Bell Telephone Directory for Jacksonville. Southern Bell Telephone, July 1941.

Southern Bell Telephone Directory for Jacksonville. Southern Bell Telephone, July 1965.

Strickland, Sandy. "Call Box." *Florida Times-Union*, August 28, 2017.

Taylor, Simon. "Lunch with Hurley Haywood." *MotorSport* (July 2016).

Teachen, John R., Jr. "Between World Wars, GM Set Franchise Pace." *Automotive News*, September 26, 2006, 18–19.

Telephone Directory for Jacksonville and Pablo Beach, Florida. American Telephone & Telegraph Company, April 1, 1915.

Warner, Bill. "Sig Haugdahl, Claude Nolan and the Jacksonville Racing Connection." *The 13th Annual Amelia Island, Concours d'Elegance Program*, March 9, 2008.

Williams, Chambers. "Jerry Hamm Chevrolet Changes Hands, Becomes Riverside Chevrolet." *Florida Times-Union*, October 18, 2016.

Wood, Wayne W. *Jacksonville's Architectural Heritage: Landmarks for the Future.* Jacksonville: University of North Florida Press, 1989.

Wood, Wayne W., with Emily Lisska and Carole Fader. *The Jacksonville Family Album: 150 Years of the Art of Photography.* Jacksonville, FL: Jacksonville Historical Society, 2005.

YP: All Things Local. N.p.: Print Media, June 2017.

INDEX

ABOUT THE AUTHOR

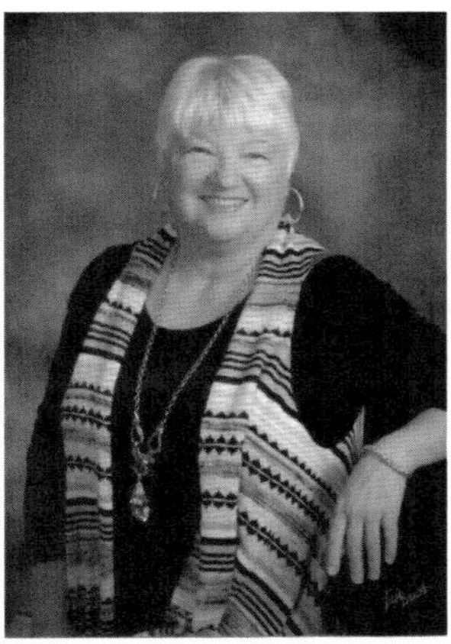

In 2007, Dorothy K. Fletcher retired after thirty-five years of teaching English in Jacksonville, Florida. At that time, she began her writing career in earnest. Already her poetry had appeared in seventy-eight literary journals, and her articles had appeared in the *Florida Times-Union*, *Jacksonville Magazine*, *Small Press Review*, the *Christian Science Monitor* and *Rider Magazine*. She also had three books published: *The Week of Dream Horses*, *The Cruelest Months* and *Zen Fishing and Other Southern Pleasures.* She began a monthly column for the *Florida Times-Union* called "By the Wayside," and that became the basis for the first of four histories she has written about her beloved city, Jacksonville, Florida. *By the Wayside* received a 2011 Jacksonville Historical Preservation Award from the Jacksonville Historic Commission of the City of Jacksonville. After that book, she wrote *Growing Up Jacksonville*, *Lost Restaurants of Jacksonville* and, most recently, *Historic Jacksonville Theatre Palaces, Drive-ins and Movie Houses*, which was a silver medal winner at the FABA Convention in August 2016. Presently, she and her husband, Hardy, reside in the San Jose area, near their children and grandchildren, and whenever they can, they go traveling.